Introduction to
Ecological Sampling

CHAPMAN & HALL/CRC
APPLIED ENVIRONMENTAL STATISTICS

Series Editors

Doug Nychka

Institute for Mathematics
Applied to Geosciences
National Center for
Atmospheric Research
Boulder, CO, USA

Richard L. Smith

Department of Statistics &
Operations Research
University of North Carolina
Chapel Hill, USA

Lance Waller

Department of Biostatistics
Rollins School of
Public Health
Emory University
Atlanta, GA, USA

Published Titles

Michael E. Ginevan and Douglas E. Splitstone, **Statistical Tools for Environmental Quality**

Timothy G. Gregoire and Harry T. Valentine, **Sampling Strategies for Natural Resources and the Environment**

Daniel Mandallaz, **Sampling Techniques for Forest Inventory**

Bryan F. J. Manly, **Statistics for Environmental Science and Management, Second Edition**

Bryan F. J. Manly and Jorge A. Navarro Alberto, **Introduction to Ecological Sampling**

Steven P. Millard and Nagaraj K. Neerchal, **Environmental Statistics with S Plus**

Wayne L. Myers and Ganapati P. Patil, **Statistical Geoinformatics for Human Environment Interface**

Éric Parent and Étienne Rivot, **Introduction to Hierarchical Bayesian Modeling for Ecological Data**

Song S. Qian, **Environmental and Ecological Statistics with R**

Thorsten Wiegand and Kirk A. Moloney, **Handbook of Spatial Point-Pattern Analysis in Ecology**

CHAPMAN & HALL/CRC
APPLIED ENVIRONMENTAL STATISTICS

Introduction to Ecological Sampling

Edited by

Bryan F. J. Manly
Western EcoSystem Technology, Inc.
Laramie, Wyoming, USA

Jorge A. Navarro Alberto
Universidad Autónoma de Yucatán
Mérida, Mexico

CRC Press
Taylor & Francis Group
Boca Raton London New York

CRC Press is an imprint of the
Taylor & Francis Group an **informa** business

A CHAPMAN & HALL BOOK

CRC Press
Taylor & Francis Group
6000 Broken Sound Parkway NW, Suite 300
Boca Raton, FL 33487-2742

© 2015 by Taylor & Francis Group, LLC
CRC Press is an imprint of Taylor & Francis Group, an Informa business

Printed on acid-free paper
Version Date: 20140603

International Standard Book Number-13: 978-1-4665-5514-3 (Hardback)

Library of Congress Cataloging-in-Publication Data

Introduction to ecological sampling / editors, Bryan F.J. Manly and Jorge A. Navarro Alberto.
    pages cm. -- (Chapman & Hall/CRC applied environmental statistics)
    Includes bibliographical references and index.
    ISBN 978-1-4665-5514-3
    1. Ecology--Statistical methods. 2. Sampling (Statistics) I. Manly, Bryan F. J., 1944- II. Navarro Alberto, Jorge A., 1963-

QH541.15.S72I58 2014
570.72'3--dc23   2014015552

Visit the Taylor & Francis Web site at
http://www.taylorandfrancis.com

and the CRC Press Web site at
http://www.crcpress.com

Contents

Preface

The origins of this book go back about 20 years when Lyman McDonald and Bryan Manly started giving courses for biologists on environmental and ecological sampling. For this purpose, they wrote notes about the different topics for use by the students in the courses and gave lectures based on the contents of the notes. These notes were subsequently used in 2006 when Bryan Manly and Jorge Navarro organized a workshop at the University of Yucatan, Mexico, directed to a biologically inclined audience. Then, in 2007 Bryan Manly was asked to set up an Internet course on environmental and ecological sampling for the Institute for Statistics Education (http://www.statistics.com). These experiences helped us realize that the range of methods covered were reachable to biologists seeking a compendium of sampling procedures for ecological and environmental studies, and their corresponding analyses, with only the necessary mathematical derivations for their comprehension. During all those years, the notes were updated, and essentially a draft version of a book was produced. This draft book was used for the Internet course until 2012, but at that stage the decision was made to publish a book not only covering standard sampling methods but also including chapters on the many recent developments in environmental and ecological sampling methods. This then required the involvement of more people in the production of what would become an edited book.

First, Jorge Navarro became a second editor for the book because of his experience in the application of statistics to many biological problems. It was then decided that the book required chapters on some specialized topics written by leading researchers on those topics. The book in its final form therefore covers standard sampling methods and analyses followed by chapters on adaptive sampling methods by Jennifer Brown, line transect sampling by Jorge Navarro and Raúl Diaz-Gamboa; removal and change-in-ratio methods by Lyman McDonald and Bryan Manly; plotless sampling by Jorge Navarro; mark-recapture sampling of closed populations by Jorge Navarro, Bryan Manly, and Roberto Barrientos; mark-recapture sampling on open populations by Bryan Manly, Jorge Navarro, and Trent McDonald; occupancy models by Darryl MacKenzie; sampling designs for environmental modeling by Trent McDonald; and trend analysis by Timothy Robinson and Jennifer Brown.

This book could not have been produced without the assistance of the seven authors who produced all or part of chapters. We are therefore grateful for their help. Finally, the contribution of Jorge Navarro as a coeditor would not have been possible without the support and resources provided by West Incorporated and the University of Wyoming during a sabbatical stay in Laramie, for which he is very grateful.

Some chapters have supplementary material that can be found at the Google site   https://sites.google.com/a/west-inc.com/introduction-to-ecological-sampling-supplementary-materials/home.

Bryan Manly
Western EcoSystems Technology Inc.
Laramie, Wyoming, USA

Jorge A. Navarro Alberto
Universidad Autónoma de Yucatán
Mérida, Yucatán, México

About the Editors

Bryan Manly was a professor of statistics at the University of Otago, Dunedin, New Zealand until 2000, after which he moved to the United States to work as a consultant for Western EcoSystems Technology. His interests are in all aspects of statistics applied to biological problems but in recent years particularly in analyses related to organisms in seas and rivers.

Jorge Navarro Alberto is a professor at the Autonomous University of Yucatan, Mexico, where, since 1986, he has taught statistics and sampling design courses for undergraduate biology students and since 1994 has taught graduate marine biology and natural resource management students. His current research involves the development of statistical methods in community ecology, biodiversity conservation, and biogeography.

Contributors

Roberto Carlos Barrientos-Medina
Universidad Autónoma de Yucatán
Departamento de Ecologia Tropical
Campus de Ciencias Biológicas y
  Agropecuarias
Mérida, México

Jennifer Brown
University of Canterbury
School of Mathematics and Statistics
Christchurch, New Zealand

Raúl Díaz-Gamboa
Universidad Autónoma de Yucatán
Departamento de Recursos Marinos
  Tropicales
Campus de Ciencias Biológicas y
  Agropecuarias
Mérida, México

Darryl MacKenzie
Proteus Wildlife Research
  Consultants
Dunedin, New Zealand

Lyman McDonald
Western EcoSystems Technology
  Inc.
Laramie, Wyoming, USA

Trent McDonald
Western EcoSystems Technology
  Inc.
Laramie, Wyoming, USA

Timothy Robinson
University of Wyoming
Department of Statistics
Laramie, Wyoming, USA

1

Introduction

Bryan Manly and Jorge Navarro

1.1 Why a Book on Ecological Sampling and Analysis?

Some biological sciences, like agronomy, physiology, and the like, share concepts and research methods with the science of ecology, but there are obvious differences regarding the way data are gathered. In the former subjects, designed experiments with the sampling processes carried out under controlled conditions are more often performed. Although the use of experiments in ecology has always been suggested in order to gain a better idea of cause-and-effect relationships (Underwood, 1997), this is often difficult or impossible, and ecologists are compelled to use observational methods of sampling. Moreover, there are multidisciplinary areas combining the ecological component and socioeconomic approaches (e.g., ethnobiology, natural resource management) in which the only sampling strategy available is a nonexperimental, observational approach.

Many ideas behind sampling in ecology have their roots in methods provided by a classic discipline in statistics, namely, survey sampling of finite populations. Simple random sampling, stratified sampling, and systematic sampling are all firsthand tools applicable to ecological studies. However, there are often particular problems faced by ecologists for sampling real animal or plant populations that have made both ecologists and statisticians develop special sampling methods that take into account the peculiarities of the situation of interest. There are numerous examples of such methods, including mark-recapture sampling, adaptive sampling, removal sampling, and so on. Our purpose in presenting this book is to cover both the classic approaches and the methods needed by working ecologists.

Although the sampling procedures covered in this book are diverse, they are unified by the widespread interest in ecological studies of estimating biological (e.g., population size and density) and environmental (e.g., the concentration of chemical elements) parameters. Thus, this book emphasizes

how particular ecological sampling methods in different settings are combined with estimation procedures that are justified by statistical theory.

1.2 The Scope and Contents of the Book

This book introduces ecological sampling methods and the analysis of the
data obtained with the assumption that readers start with basic knowledge
of standard statistical methods based on one or two introductory courses but
know little about how these methods are applied with ecological data and
know nothing about the more specialized methods that have been developed specifically for ecological data. The book is only an introduction, so
that use of many of the methods described in the chapters may require read-
ing a text that is more specialized or in some cases even attending a course
on the use of a special statistical computer package.

There are ten chapters in the rest of the book. The remainder of this chapter briefly describes what is in these chapters and the reason why each chapter was included in the book.

Chapter 2, "Standard Sampling Methods and Analyses," covers the traditional methods that have been employed with ecological and other data for
more than 50 years, plus some newer developments in this area. These methods all assume that there is a specific population of interest, and that the population consists of many items of interest. For example, the population could
be all of the plants of a certain species in a national park, and the interest is
in the number of plants per square meter for the entire park. In these types of
situations, it is usually the case that obtaining information for the whole population is not possible because this would be far too expensive. Therefore, it
is necessary to sample the population and estimate variables of interest using
the sample results. With the plant population, this could be done by sampling
meter-square plots at random throughout the national park and estimating
the density for the whole park using the observed density on the sampled
plots, with a measure of how large the difference between the observed density and the true density is likely to be. Chapter 2 discusses estimation using
random sampling methods like this and a number of variations on this that
are intended to improve the estimation, such as stratified sampling, by which
different parts of the population are sampled separately.

Chapter 3, "Adaptive Sampling Methods," covers the methods that have
been proposed for which there is an initial sampling of a population and the
results from this are used to decide where to sample after that, with the idea
that this should lead to more efficient sampling. Adaptive cluster sampling,
which is covered in more detail than other adaptive sampling methods,
involves dividing an area of interest into quadrats, taking a random sample
of those quadrats, and then taking more samples adjacent to the quadrats

where the items of interest are either present or the number exceeds some threshold. Similarly, two-phase adaptive sampling involves taking samples from different geographical parts of a population (strata) and then deciding where to do more sampling based on what is found in the initial sampling. The idea is then that the second round of sampling can be in those parts of the population where more sampling will be most beneficial. These and the other methods considered in Chapter 3 are generally concerned with estimating either the total abundance or density per unit area of animals or plants in a region.

Chapter 4, "Line Transect Sampling," is mainly concerned with the situation when one or more observers travel along a line in a region of interest and record the number of objects of interest seen (again animals or plants) and the distance of these objects from the line. Typically, this may involve following a line on the ground, flying in an aircraft and observing objects on the ground, or moving along a path in a boat over a sea or lake and recording the objects seen. It is generally assumed that the probability of detecting an object depends on its distance from the line, with a low probability of detection for objects far away. The data are then used to estimate a detection function and then estimate the total number of objects within a certain distance from the line and the density of objects per unit area around the line. This method of sampling is usually used when alternative methods such as randomly sampling quadrats in the region of interest are not practical for some reason.

Chapter 5, "Removal and Change-in-Ratio Methods," discusses two methods for estimating the size of animal populations. With the removal method, an animal population is sampled a number of times, and the animals caught are removed or possibly marked so that if they are caught again it will be known that they were seen before so that they are effectively "removed." Assuming a closed population (i.e., a population in which animals are not entering and leaving between samples), the number of animals left in the population will decrease every time a sample is taken, so that the number of animals available for removal will decrease with time and will be zero if enough samples are taken. At that time, the population size will consist of the total number of animals removed in the earlier samples. However, it is not necessary to remove all of the animals to obtain an estimate of the total population size; Chapter 5 describes how the results from several samples can be used to estimate the total number of animals in a population even though some of these animals have not been seen. The change in ratio method is similar but involves a population in which there are two or more recognizable types of animal in a population, such as males and females or juveniles and adults. Then, a sample is taken, and a fixed number of one of the animal types is removed or possibly marked so that a marked animal is considered removed. If a second sample is taken, then it becomes possible to estimate the population sizes at the time of the first sample and at the time of the second sample (which is just the first sample size minus the number of animals removed). Both the removal method and the change in ratio method are closely related to the mark-recapture methods described in Chapter 7.

Chapter 6, "Plotless Sampling," is concerned with methods for estimating the density of objects, such as the trees in a defined area, without dividing the area into plots of a certain size and then randomly sampling the plots or sampling the area by one of the other methods discussed in Chapters 2 to 5. Instead, points are selected in the area of interest either randomly or with a systematic pattern, then the distance to the nearest object is measured for each point and possibly the distance from that object to its nearest or *k*th-nearest neighbor. There are various methods that have been proposed for that type of sampling, and Chapter 6 describes two of these in some detail. The first is T-square sampling. In this case, once a point in the area of interest is selected for sampling, then the distance from this point to the nearest object is measured. A line is then set passing through the position of the object at a right angle to the line from the initial point to the object. The distance from the object to its nearest neighbor on the side of the line away from the initial point is then measured. Based on the two measured distances (from the point to the first object and the first object to its nearest neighbor) for a number of points in the area of interest, it is possible to estimate the density of objects in the whole area, assuming that the location of objects is at least approximately random in the area. The second method described in Chapter 6 is wandering quarter sampling. This starts with a randomly selected point in the region of interest. A direction is then selected, such as west, and the nearest object within the 90° angle from southwest to northwest is found. The distance from this object to the next object in the southwest-to-northwest direction is then measured. This procedure is continued until *n* distances have been measured and it is possible to estimate the density of points in the sampled area based on these distances.

Chapter 7, "Introduction to Mark-Recapture Sampling and Closed-Population Models," covers these methods for closed populations (with no losses or gains of animals during the sampling period). With closed population methods, there is a first sample time when captured animals are marked and released, and then one or more further samples are taken, with captured animals again marked and released. Then, it is possible to estimate the population size subject to certain assumptions about the capture process.

Chapter 8, "Open-Population Mark-Recapture Models," covers situations when the population size can change between sample times because of losses (deaths and emigration) and gains (births and immigration). With open populations, the sampling process is similar to that for closed populations, but there are generally more samples. It becomes possible to estimate the population size at the sample times other than the first and last, the survival rates between sample times except between the last two samples, and the number of new entries to the population between the sample times except between the last two samples. This is again subject to certain assumptions about the capture process and the survival rates and the new entries to the population.

Chapter 8 describes the traditional methods for analyzing data and developments that are more recent. In addition, methods for analyzing data based on the recovery of dead animals are considered.

Chapter 9, "Occupancy Models," covers situations where there are data on the recorded presences and absences of species in different locations but some of the absences might be because the species was present but not detected. This then leads to the idea of having a model for the probability that a species is present at a location based on its characteristics and a model for the probability of detecting a species if it is present, again based on the characteristics of the location. These models then require more than one sample to be taken from each location because if a location is sampled several times and the species is seen at least once, then this provides information on the probability of detecting a species when it is really present. Although site occupancy models were originally just used when there were presences and absences, they have now been extended to situations with more than two possibilities regarding presence, such as absence, presence with breeding, and presence without breeding. Chapter 9 therefore also considers these types of situations and situations where the status of a location may change with time.

Chapter 10, "Sampling Designs Used for Environmental Monitoring," describes the various types of sampling schemes that are being used for studies to track changes in environmental variables at local, national, and international levels. It notes the difference between these types of studies and research studies that are usually carried out over a shorter period of time to examine possible changes related to an issue of concern (e.g., a study to measure the adverse effects of an oil spill). The chapter considers the different types of spatial designs used with environmental monitoring and the possible designs that allow for repeated sampling at individual locations and changes with time of these locations.

Finally, Chapter 11, "Models for Trend Analysis," considers in detail the types of analyses that are appropriate when there are repeated observations over time at a number of sampling sites and there is interest in both changes at the individual sites and changes for the whole geographical area covered by the sites. The chapter considers the use of various simple analyses and graphical methods for exploring trends in the data and two approaches for modeling. One approach involves using linear regression methods for examining the trends at individual sampling sites and then combining these results to examine the overall trend for all sites together. The other approach does one analysis using the data for all of the sites together based on what is called a mixed model, which essentially allows for individual sites to display random differences from an overall trend. This chapter uses a data set on mercury concentrations in fish to illustrate the different methods considered, with the data collected from the yearly sampling of fish from the same 10 randomly chosen locations in a lake for 12 years.

2

Standard Sampling Methods and Analyses

Bryan Manly

2.1 Introduction

Often, the goal of ecological sampling is to summarize the characteristics of the individual units in a biological population. For example, the characteristics of interest might be the weight and sex of the individual animals in a population in a particular area. A summary of these characteristics then consists of estimates of the mean and standard deviation of the weight of animals and an estimate of the proportion of females.

In the terminology of statistics, the *population* is defined to be the collection of all items that are of interest in an investigation. Thus, the items might be individual animals or plants, but they could also be small plots of land, pieces of rock, or groups of animals. As far as statistical theory is concerned, it is crucial that the items that make up the population are sampled using an appropriate procedure. For this reason, the items are often called the sample units. Sometimes, population sizes are small enough to allow every item to be examined. This then provides a census. However, the populations of interest in ecology are usually large enough to make a census impractical.

The measures that are used to summarize a population are called population parameters, and the corresponding sample values are called statistics. For example, the population mean (a parameter) might be estimated by a sample mean (a statistic). Similarly, a population proportion of females (a parameter) might be estimated by a sample proportion (a statistic).

2.2 Simple Random Sampling

The use of random sampling is important whenever inferences are to be made about population parameters on the basis of the sample result because of the need to base the inferences on the laws of probability. In this connection, it must be appreciated that random sampling is not the same as a haphazard selection of the units to be surveyed. Rather, it involves the selection of units using a well-defined and carefully carried out randomization procedure that (in a simple application) ensures that all possible samples of the required size are equally likely to be chosen.

Although this is the case, it is a fact that with ecological sampling the strictly random selection of sample units is not always possible. This issue is discussed further in this chapter and in other chapters as well. For the moment, it is assumed that true random sampling can be carried out.

Because all possible samples can occur with random sampling, it is obvious that this method might produce exactly the same units as a haphazardly drawn sample. It is important therefore to appreciate that the key to the value of random sampling is the properties of the sampling procedure rather than the specific units that are obtained. In fact, it is not uncommon to feel uncomfortable with the result of random sampling because it does not look representative enough. But, this will not be a valid objection to the sample providing that the procedure used to select it was defined and carried out in an appropriate manner.

Simple random sampling involves giving each sample unit the same probability of selection. This can be with replacement, in which case every selected unit is chosen from the full population irrespective of which units have already been included in the sample, or without replacement, in which case a sample unit can occur at most once in the sample. As a general principle, sampling without replacement is preferable to sampling with replacement because it gives slightly more accurate estimation of population parameters. However, the difference between the two methods of sampling is not great when the population size is much larger than the sample size.

EXAMPLE 2.1 Sampling Plants in a Large Study Area

Suppose that it is required to estimate the density of plants of a certain species in a large study area. Then, one approach would be to set up a grid and consider the area to consist of the quadrats that this produces. Figure 2.1 indicates the type of result that might then be obtained; in this case, there are 120 quadrats covering a rectangular study area. The quadrats are then the sampling units that make up the population of interest. The list of these units is sometimes called the sampling frame.

The next step would be to decide on a sample size n; that is, how many quadrats should be randomly sampled to estimate the population mean number of plants per quadrat with an acceptable level of accuracy? Methods for choosing sample sizes are discussed further elsewhere, but for this example, it is assumed that a sample size of 12 is needed.

There are various ways to determine the random sample. It is definitely not allowable just to think up the numbers because human beings do not have random number generators in their heads. One approach in

1	2	3	4	5	6	7	8	9	10
11	12	13	14	15	16	17	18	19	20
21	22	23	24	25	26	27	28	29	30
31	32	33	34	35	36	37	38	39	40
41	42	43	44	45	46	47	48	49	50
51	52	53	54	55	56	57	58	59	60
61	62	63	64	65	66	67	68	69	70
71	72	73	74	75	76	77	78	79	80
81	82	83	84	85	86	87	88	89	90
91	92	93	94	95	96	97	98	99	100
101	102	103	104	105	106	107	108	109	110
111	112	113	114	115	116	117	118	119	120

FIGURE 2.1
A rectangular study area divided into 120 quadrats to be used as sample units. The numbers shown in the quadrats are labels used for random sampling rather than the values of some variable of interest.

the present case involves labeling the quadrats in the population from 1 to 120 (as in Figure 2.1) and then selecting the ones to sample by generating 12 random integers from 1 to 120 on a computer. If R is a computer-generated random number in the range 0 to 1, then $Z = MIN[120, INT(R \times 120 + 1.0)]$ is a random integer in the range 1 to 120, where the function $INT(x)$ gives the integer part of x, and $MIN(a,b)$ gives the minimum of a and b. The minimum function is included here in case a value of $R = 1$ can occur, in which case $INT(R \times 120 + 1.0) = 121$.

Because sampling should be without replacement, the same quadrat would not be allowed to occur more than once. Any repeated selections would therefore be ignored and the process of selecting random integers continued until 12 different quadrats have been chosen.

A table of random numbers such as Table 2.1 can also be used for the selection of sample units. To use this table, first start at an arbitrary place in the table such as the beginning of row 8. The first three digits in each block of four digits can then be considered, to give the series 569, 362, 898, 287, 607, 099, 681, 779, 458, 883, 181, 001, 927, 280, 224, 831, 711, 207, 151, 180, 978, 773, 075, 367, 251, 106, 547, 711, 347, 720, 737, and so on. The first 12 different numbers between 1 and 120 then give a simple random sample of quadrats, that is, 99, 1, 75, and so on.

Once the sample quadrats are chosen, these would be examined to find the number of plants that each contains. The average for the 12 sample quadrats then gives an estimate of the mean number of plants in the area of 1 quadrat over the entire study region, for which the likely level of error can be determined by methods discussed next. If necessary, the estimate and its error can then be converted so it is in terms of plants per square meter or any other measure of density that is of interest.

TABLE 2.1

A Random Number Table with Each Digit Chosen Such That 0, 1, . . . , 9 Were
Equally Likely to Occur

1252	9045	1286	2235	6289	5542	2965	1219	7088	1533
9135	3824	8483	1617	0990	4547	9454	9266	9223	9662
8377	5968	0088	9813	4019	1597	2294	8177	5720	8526
3789	9509	1107	7492	7178	7485	6866	0353	8133	7247
6988	4191	0083	1273	1061	6058	8433	3782	4627	9535
7458	7394	0804	6410	7771	9514	1689	2248	7654	1608
2136	8184	0033	1742	9116	6480	4081	6121	9399	2601
5693	3627	8980	2877	6078	0993	6817	7790	4589	8833
1813	0018	9270	2802	2245	8313	7113	2074	1510	1802
9787	7735	0752	3671	2519	1063	5471	7114	3477	7203
7379	6355	4738	8695	6987	9312	5261	3915	4060	5020
8763	8141	4588	0345	6854	4575	5940	1427	8757	5221
6605	3563	6829	2171	8121	5723	3901	0456	8691	9649
8154	6617	3825	2320	0476	4355	7690	9987	2757	3871
5855	0345	0029	6323	0493	8556	6810	7981	8007	3433
7172	6273	6400	7392	4880	2917	9748	6690	0147	6744
7780	3051	6052	6389	0957	7744	5265	7623	5189	0917
7289	8817	9973	7058	2621	7637	1791	1904	8467	0318
9133	5493	2280	9064	6427	2426	9685	3109	8222	0136
1035	4738	9748	6313	1589	0097	7292	6264	7563	2146
5482	8213	2366	1834	9971	2467	5843	1570	5818	4827
7947	2968	3840	9873	0330	1909	4348	4157	6470	5028
6426	2413	9559	2008	7485	0321	5106	0967	6471	5151
8382	7446	9142	2006	4643	8984	6677	8596	7477	3682
1948	6713	2204	9931	8202	9055	0820	6296	6570	0438
3250	5110	7397	3638	1794	2059	2771	4461	2018	4981
8445	1259	5679	4109	4010	2484	1495	3704	8936	1270
1933	6213	9774	1158	1659	6400	8525	6531	4712	6738
7368	9021	1251	3162	0646	2380	1446	2573	5018	1051
9772	1664	6687	4493	1932	6164	5882	0672	8492	1277
0868	9041	0735	1319	9096	6458	1659	1224	2968	9657
3658	6429	1186	0768	0484	1996	0338	4044	8415	1906
3117	6575	1925	6232	3495	4706	3533	7630	5570	9400
7572	1054	6902	2256	0003	2189	1569	1272	2592	0912
3526	1092	4235	0755	3173	1446	6311	3243	7053	7094
2597	8181	8560	6492	1451	1325	7247	1535	8773	0009
4666	0581	2433	9756	6818	1746	1273	1105	1919	0986
5905	5680	2503	0569	1642	3789	8234	4337	2705	6416
3890	0286	9414	9485	6629	4167	2517	9717	2582	8480
3891	5768	9601	3765	9627	6064	7097	2654	2456	3028

2.3 Estimation of Mean Values

Assume that a simple random sample of size n is taken from a population of N units, and that the variable of interest Y has values $y_1, y_2, \ldots, y_n$ for the sampled units. Then, sample statistics that are commonly computed are the sample mean

$$\bar{y} = (y_1 + y_2 \ldots + y_n)/n = \left\{ \sum_{i=1}^{n} y_i \right\}/n, \tag{2.1}$$

the sample variance

$$s^2 = \left\{ \sum_{i=1}^{n} (y_i - \bar{y})^2 \right\}/(n-1), \tag{2.2}$$

where s is the sample standard deviation (the square root of the variance), and $\widehat{CV}(y) = s/\bar{y}$ is the estimated coefficient of variation. Note the use of a caret ^ to indicate a sample estimate. This is a common convention in statistics that is used frequently in this chapter.

The coefficient of variation is often just referred to as the CV. Also, it is often expressed as a percentage because $100s/\bar{y}$ is the standard deviation as a percentage of the mean.

The sample mean is an estimator of the population mean μ, where this is the mean of Y for all N units in the population. The difference $\bar{y} - \mu$ is the sampling error; this will vary from sample to sample if the sampling process is repeated. It can be shown theoretically that if the random sampling process is repeated many times, then the sampling error will average out to zero. Therefore, the sample mean is an unbiased estimator of the population mean.

It can also be shown theoretically that the distribution of $\bar{y}$ that is obtained by repeating the process of simple random sampling has variance

$$\mathrm{Var}(\bar{y}) = \{\sigma^2/n\}\{1 - n/N\}, \tag{2.3}$$

where σ^2 is the variance of the Y values for the N units in the population. In this equation, the factor $\{1 - n/N\}$ is called the finite population correction. The square root of $\mathrm{Var}(\bar{y})$ is commonly called the standard error of the sample mean, which is denoted by $\mathrm{SE}(\bar{y})$.

Equation (2.3) might not be familiar to those who have taken a standard introductory course in statistics because it is usual in such courses to assume that the population size N is infinite, leading to $n/N = 0$ and $\mathrm{Var}(\bar{y}) = \sigma^2/n$.

Equation (2.3) is therefore more general than the equation that is often quoted for the variance of a sample mean.

The variance of the sample mean can be estimated by

$$\widehat{\text{Var}}(\bar{y}) = \{s^2/n\}\{1 - n/N\}. \tag{2.4}$$

The square root of this quantity is the estimated standard error of the mean

$$\widehat{\text{SE}}(\bar{y}) = \sqrt{\left[\{s^2/n\}\{1 - n/N\}\right]}, \tag{2.5}$$

and the estimated CV of the mean is $\widehat{\text{CV}}(\bar{y}) = \widehat{\text{SE}}(\bar{y})/\bar{y}$.

The terms *standard error of the mean* and *standard* deviation are often confused when encountered for the first time. What must be remembered is that the standard error of the mean is the standard deviation of the mean rather than the standard deviation of individual observations. More generally, the term *standard error* is used to describe the standard deviation of any sample statistic that is used to estimate a population parameter.

The CV of the mean is an index that reflects the precision of estimation relative to the magnitude of the mean. This can be used to compare the results of several studies to see which have relatively better precision than others. Often, this is used to define the required precision of a study. For example, in estimating the size of the shellfish of a certain species on a beach, it might be required that the population size be estimated with a CV of less than 20%. In practice, this would mean that after the sampling has taken place, it should be found that $100\,\widehat{\text{SE}}(\bar{y})/\bar{y} < 20$.

As an example of the calculation of the statistics that have just been defined, suppose that a random sample of size $n = 5$ is taken from a population of size $N = 100$ and the sample values are found to be 1, 4, 3, 5, and 8. Then, $\bar{y} = 4.20$ and $s^2 = 6.70$, so that $s = \sqrt{6.70} = 2.59$, $\widehat{\text{CV}}(y) = 2.59/4.20 = 0.62$, $\widehat{\text{Var}}(\bar{y}) = \{6.70/5\}$ $\{1 - 5/100\} = 1.27$, $\widehat{\text{SE}}(\bar{y}) = \sqrt{1.27} = 1.13$, and $\widehat{\text{CV}}(\bar{y}) = 1.13/4.20 = 0.27$. These calculations have been carried through to two decimal places. As a general rule, it is reasonable for statistics to be quoted with at least one more decimal place than the original data, which suggests that results should be given to at least one decimal place in this example.

The accuracy of a sample mean for estimating the population mean is often represented by a $100(1 - \alpha)\%$ confidence interval of the form

$$\bar{y} \pm z_{\alpha/2}\,\widehat{\text{SE}}(\bar{y}), \tag{2.6}$$

where $z_{\alpha/2}$ refers to the value that is exceeded with probability $\alpha/2$ for the standard normal distribution. Common values used for the confidence level with the corresponding values of z are $z_{0.25} = 0.68$ for 50% confidence, $z_{0.16} = 1.00$ for 68% confidence, $z_{0.05} = 1.64$ for 90% confidence, $z_{0.025} = 1.96$ for

95% confidence, and $z_{0.005} = 2.58$ for 99% confidence. The meaning of the confidence interval is best understood in terms of a specific case. For instance, the interval $\bar{y} \pm 1.64\widehat{SE}(\bar{y})$ will contain the true population mean with a probability of approximately 0.90. In other words, about 90% of the intervals calculated in this way will contain the true population mean.

The interval (2.6) is only valid for large samples. For small samples (say with $n < 20$), it will be better to use

$$\bar{y} \pm t_{\alpha/2, n-1} \cdot \widehat{SE}(\bar{y}),\qquad(2.7)$$

where $t_{\alpha/2, n-1}$ is the value that is exceeded with probability $\alpha/2$ for the t distribution with $n - 1$ degrees of freedom. This requires the assumption that the variable being measured is approximately normally distributed in the population being sampled. If this is not the case, then no simple method exists for calculating an exact confidence interval.

2.4 Estimation of Totals

In many situations, the ecologist is more interested in the total of all values in a population rather than the mean per sample unit. For example, the total weight of new growth of all the plants in a region might be more important than the mean growth on individual plants. Similarly, the total amount of forage eaten by a herd of animals might be more important than the average amount eaten per animal.

The estimation of a population total is straightforward if the population size N is known and an estimate of the mean is available. It is obvious, for example, that if each of 500 animals is estimated to require an average of 25 kg of forage, then $500 \times 25 = 12{,}500$ kg of forage is required for all animals. The general equation that applies is that the estimated population total for the variable Y is

$$\hat{T}_y = N\bar{y},\qquad(2.8)$$

the mean per unit multiplied by the number of units. The sampling variance of $\hat{T}_y$ is then

$$\mathrm{Var}(\hat{T}_y) = N^2 \mathrm{Var}(\bar{y}),\qquad(2.9)$$

and its standard error (i.e., standard deviation) is

$$\mathrm{SE}(\hat{T}_y) = N\, \mathrm{SE}(\bar{y}),\qquad(2.10)$$

so that estimates of the variance and standard error are $\widehat{\text{Var}}(\bar{T}_y) = N^2 \widehat{\text{Var}}(\bar{y})$, and $\widehat{\text{SE}}(\bar{T}_y) = N\widehat{\text{SE}}(\bar{y})$.

An approximate $100(1 - \alpha)\%$ confidence interval for the true population total can also be calculated in essentially the same manner as described in the previous section for finding a confidence interval for the population mean. For example, an approximate 95% confidence interval is

$$\hat{T}_y \pm 1.96\, \widehat{\text{SE}}(\bar{T}_y). \tag{2.11}$$

EXAMPLE 2.2 Estimation of a Total by Strip Transect Sampling

Suppose that there is interest in using strip transect sampling for estimating the number of deer pellet groups in a study area. Assume that to this end the area is covered with 5808 strip transects each with a width of 3 m and a length 20 m, and 20 of these are randomly sampled without replacement. The number of pellet groups in each sample transect is then determined by careful searching.

If this procedure yields a mean of $\bar{y} = 5.55$ pellet groups per transect with a standard deviation of $s = 3.75$, then the estimated standard error of $\bar{y}$ is found to be $\widehat{\text{SE}}(\bar{y}) = \sqrt{[\{3.75^2/20\}\{1 - 20/5808\}]} = 0.837$, and the estimated number of pellet groups in the entire study area is $\hat{T}_y = N\bar{y} = 5808 \times 5.55 = 32234.4$, with an estimated standard error of $\widehat{\text{SE}}(\bar{T}_y) = N\widehat{\text{SE}}(\bar{y}) = 5808 \times 0.837 = 4861.3$. Finally, an approximate 95% confidence interval for the true total number of pellet groups is $32{,}234.4 \pm 1.96 \times 4861.3$, which gives the limits 22,706 to 41,763 rounded to the nearest integers.

2.5 Sample Sizes for Estimation of Means

One of the key considerations in designing a study is the sample size that will be used. This should be large enough to give adequate accuracy for the estimation of the population parameters of interest but should not be unnecessarily large. The sample size is determined by the resources available and the properties desired for the estimates, but in most studies the resources available have the greatest influence. In some cases, the resources available do not allow reasonably accurate estimation. In that case, serious consideration needs to be made about whether the study should proceed or whether there is some alternative way to obtain the desired information about the parameters of interest.

There are various approaches for determining an appropriate sample size for a study. For example, it might be decided that the 95% confidence interval for the mean that is obtained should be $\bar{y} \pm d$, where d is some suitable small value. This then represents an absolute level of precision. Alternatively, the accuracy might be specified in terms of the CV of the mean, $CV(\bar{y}) = 100\,SE(\bar{y})/\mu$, so that, for example, $CV(\bar{y})$ should be no more than 20%. This then represents the precision relative to the population mean. The CV approach is particularly useful when a number of different quantities are being estimated and the true population means may vary considerably.

To obtain a $100(1-\alpha)\%$ confidence interval for the mean of $\bar{y} \pm d$, it is required from Equation (2.6) that $z_{\alpha/2}\,\widehat{SE}(\bar{y}) = d$, so that $z_{\alpha/2}\{s/\sqrt{n}\}\{1-n/N\} = d$. Solving this equation for n yields

$$n = z_{\alpha/2}^2 s^2/(d^2 + z_{\alpha/2}^2 s^2/N). \qquad (2.12)$$

Because the standard deviation s is not known in advance, it is necessary to guess what this might be, for example, using the value from a previous sample.

If the population size N is large, then Equation (2.12) simplifies to

$$n \approx (z_{\alpha/2}s/d)^2. \qquad (2.13)$$

This is a conservative equation in the sense that, for all population sizes N, Equation (2.13) gives a larger value of n than Equation (2.12).

If a specific CV is desired rather than a specific confidence interval for the mean, then this requires that, for the sample obtained, $\widehat{SE}(\bar{y})/\bar{y} = r$, where r is the required CV. Then, squaring both sides gives $\{s^2/n\}\{1-n/N\}/\bar{y}^2 = r^2$, so that $n = (s/\bar{y})^2/\{r^2 + (s/\bar{y})^2/N\}$, or

$$n = \widehat{CV}(y)^2/\{r^2 + \widehat{CV}(y)^2/N\}. \qquad (2.14)$$

This equation can be applied with an estimated or guessed value for the CV of the population in place of the unknown $\widehat{CV}(y)$. For a large population size N, it reduces to

$$n = \{\widehat{CV}(y)/r\}^2, \qquad (2.15)$$

which always gives a larger value than Equation (2.14).

Equations (2.12) to (2.15) should be regarded as giving no more than a rough indication of adequate sample sizes when they are used with guessed values for standard deviations or CVs. Nevertheless, as a general principle it is true that any effort spent in determining appropriate sample sizes is better than no effort at all.

All of the discussion so far about sample sizes has been in terms of just one variable of interest in a study, but in most studies, several different variables have to be considered at the same time. If all variables require sample sizes of about the same magnitude, then the size used can be the maximum required for any variable. However, if this is not possible within the available resources, then some smaller size might need to be used, and some variables might have to be estimated with less precision than was originally desired.

EXAMPLE 2.3 Determining the Sample Size for Strip Transect Sampling

In a pilot study of winter mortality of deer in central Wyoming, strip transects 1 km long by 60 m wide were walked to look for dead animals. In 36 randomly selected transects, 12 dead deer were counted to give a sample mean of $\bar{y} = 12/36 = 0.333$ dead deer per transect, with a sample standard deviation of $s = 0.828$. The estimated percentage CV of counts per transect was therefore $CV(y) = 100 \times 0.828/0.333 = 248.6\%$.

Suppose that it was determined that the target CV for estimating the population mean number of dead deer per transect for the main study was 15%. There were about $N = 50,000$ possible transects, so Equation (2.14) shows that the number of transects needed to be sampled to reach this target is $n = 2.486^2/\{0.15^2 + 2.486^2/50,000\} = 273.2$ transects, say 273 transects. Because the population size N is so large, Equation (2.15) gives almost the same result, with $n = 2.486^2/0.15^2 = 274.7$.

2.6 Errors in Sample Surveys

In general, there are four sources of error or variation in scientific studies (Cochran, 1977). First, the observations vary with the sample units, and as a result, different random samples will generally produce different estimates of population parameters. This variation is just because of sampling errors. Second, there may be errors caused by the lack of uniformity in the manner in which a study is conducted. The measurement procedure might be biased, imprecise, or both biased and imprecise. This type of measurement error results solely from the manner in which the observations are made. For example, a fisher might report incorrect lengths and weights of the fish caught, human subjects might lie about their age or weight, or a measurement instrument might not be correctly calibrated. Third, there might be missing data because of the failure to measure some units in the sample. This will introduce a bias if the missing values are unusual in some way. For example, in a study of vegetation, some sample plots might be inaccessible, and in fact these plots have high densities of the plants of interest. Finally, errors might be introduced in recording, typing, and editing data.

An understanding of sampling errors and their effects is the basis of statistical inference procedures, with the assumption that these errors are far more important than the other errors. In reality, however, the other three types of error might be more important than the sampling errors unless great care is taken with the study design, with well-documented and controlled sampling protocols. This is of particular importance when the sampling for a study is carried out by a number of different people. They need careful training in all aspects of the data collection, with an emphasis on ensuring that everyone uses consistent sampling and measurement methods.

2.7 Estimation of Population Proportions

In a common situation, it is necessary to estimate the proportion, p say, of units in a population that have a particular characteristic. For example, if the units are trees, then there might be interest in the proportion of trees in a particular size class. In this situation, the population proportion can be estimated by the proportion observed in a simple random sample.

Let r denote the number of units with the characteristic of interest in a random sample of n. Then, the sample proportion is $\hat{p} = r/n$, and it can be shown that this has a sampling variance of

$$\text{Var}(\hat{p}) = \{p(1-p)/n\}\{1-n/N\}, \tag{2.16}$$

and therefore a standard error (standard deviation) of

$$\text{SE}(\hat{p}) = \sqrt{\left[\{p(1-p)/n\}\{1-n/N\}\right]}. \tag{2.17}$$

Equations (2.16) and (2.17) include the finite population correction factor $(1 - n/N)$, where N is the size of the population being sampled. If N is large relative to n or if N is unknown, then this factor is usually set equal to 1.

Estimated values for the variance and standard error can be obtained by replacing the population proportion in Equations (2.16) and (2.17) with the sample proportion $\hat{p}$. Thus,

$$\widehat{\text{SE}}(\hat{p}) = \sqrt{\left[\{\hat{p}(1-\hat{p})/n\}\{1-n/N\}\right]}. \tag{2.18}$$

This creates little error unless the sample size is quite small (say, less than 20). Using this estimate, an approximate $100(1 - \alpha)\%$ confidence interval for the true proportion is

$$\hat{p} \pm z_{\alpha/2}\widehat{\text{SE}}(\hat{p}), \tag{2.19}$$

where, as before, $z_{\alpha/2}$ is the value from the standard normal distribution that is exceeded with probability $\alpha/2$.

The confidence limits produced by Equation (2.19) are based on the assumption that the sample proportion is approximately normally distributed. A useful rule of thumb is that if $np(1 - p) \geq 5$, then this assumption is reasonable. If this is not the case, then alternative methods for calculating confidence limits should be used, as discussed, for example, by Dixon and Massey (1983).

> **EXAMPLE 2.4 Estimating the Proportion of Barren Sage Grouse Hens**
>
> A survey of sage grouse hens in an area in the state of Wyoming, USA, found that of $n = 120$ randomly sampled hens, $r = 39$ were barren. This gives an estimated proportion of $\hat{p} = 39/120 = 0.325$ barren hens in the population. Assuming that the population size N is large relative to the sample size of 120, Equation (2.18) then gives the estimated standard error of the proportion as $\widehat{SE}(\hat{p}) = \sqrt{[0.325(1-0.325)/120} = 0.0428$, and an approximate 95% confidence interval for the population proportion of barren hens is therefore $0.325 \pm 1.96 \times 0.0428$ or 0.241 to 0.408.
>
> In a random sample of 88 sage grouse hens taken in another area in Wyoming, it was found that 15 were barren, so that $\hat{p} = 15/88 = 0.170$, with $\widehat{SE}(\hat{p}) = \sqrt{[0.170(1-0.170)/88} = 0.0400$, again assuming that the population size is much larger than the sample size and therefore omitting the finite population correction. A 95% confidence interval for the proportion of barren hens in the second area is therefore $0.17 \pm 1.96 \times 0.0400$ or 0.092 to 0.248.
>
> A rough-and-ready procedure to compare two sample proportions involves checking whether the confidence intervals overlap. Here, they overlap slightly, so it appears that the proportion of barren hens might not have differed for the two districts. However, a more accurate comparison can be made by noting that the variance of the difference between the proportions in two independent samples is the sum of the individual variances, so that $Var(\hat{p}_1 - \hat{p}_2) = Var(\hat{p}_1) + Var(\hat{p}_2)$. Therefore, the difference of 0.155 between the first proportion (0.325) and the second proportion (0.170) is an estimate of the population difference with an estimated standard error of $\sqrt{\{0.0428^2 + 0.0400^2\}} = 0.0586$. On this basis, a 95% confidence interval for the true population difference is $0.155 \pm 1.96 \times 0.0586$ or 0.040 to 0.269. As this interval does not include zero, there is clear evidence that the proportion of barren hens was different for the two districts.

2.8 Determining Sample Sizes for the Estimation of Proportions

An approximate $100(1 - \alpha)$% confidence interval for a proportion based on a sample size of n from a population of size N takes the form

$\hat{p} \pm z_{\alpha/2}\sqrt{\{\hat{p}(1-\hat{p})/n\}\{1-n/N\}}$. Hence, a confidence interval of $\hat{p} \pm d$ requires that $d = z_{\alpha/2}\sqrt{[\{\hat{p}(1-\hat{p})/n\}\{1-n/N\}}$. Solving for n then gives

$$n = z_{\alpha/2}^2 \hat{p}(1-\hat{p})/\{d^2 + z_{\alpha/2}^2 \hat{p}(1-\hat{p})/N\}. \tag{2.20}$$

To use this equation, a likely value for $\hat{p}$ is guessed or a value from a previous sample is used. If N is very large or is unknown, then the second term in the denominator can be ignored and the equation simplifies to

$$n = z_{\alpha/2}^2 \hat{p}(1-\hat{p})/d^2. \tag{2.21}$$

If there are no prior data and it is not possible to guess what $\hat{p}$ might be, then the worst possible case can be assumed, which is that $\hat{p} = 0.5$. Substituting this value into Equation (2.20) gives a sample size that will be larger than necessary unless $\hat{p}$ does happen to equal 0.5.

EXAMPLE 2.5 Determining the Sample Size for Sampling Sage Grouse Hens

To illustrate the use of Equation (2.20), consider again the survey of female sage grouse hens in an area in Wyoming discussed previously. Suppose that it is desirable to estimate the proportion of barren hens such that a 90% confidence interval will consist of the sample proportion ± 0.05. The estimate from the first of two surveys was $\hat{p} = 0.325$, and it will be assumed that there are about 1000 hens in the area that was sampled. Substitution into Equation (2.20) gives

$n = 1.64^2 \times 0.325(1 - 0.325)/\{0.05^2 + 1.64^2 \times 0.325(1 - 0.325)/1000\} = 190.9$,

say, 191. This is the appropriate sample size. In a practical application, this would probably be rounded to $n = 200$ to give an estimate that is likely to be slightly better than what is required.

2.9 Stratified Random Sampling

It can be argued that simple random sampling leaves too much to chance, particularly when the sample size is small. It might, for example, be clear that the number of sampled units in different geographical areas does not match the population sizes in those areas, with parts of the population undersampled and other parts oversampled. One way to overcome this potential problem, while keeping the advantages of random sampling, is to use stratified random sampling. To this end, the units in the population are divided into nonoverlapping strata, and an independent simple random sample is selected from each of these strata.

Often, there is nothing to lose by using this more complicated type of sampling, but there are some potential gains. First, if the individuals within strata are rather more similar than individuals in general, then the estimate of the overall population mean will have a smaller standard error than can be obtained with the same simple random sample size. Second, there may be value in having separate estimates of population parameters for the different strata. Third, stratification makes it possible to sample different parts of a population in different ways, which may make some cost savings possible.

On the other hand, a stratified sampling design has problems when there are errors in allocating sample units to the strata. This may occur, for example, if the allocation is made using a map that is not completely accurate. Then, when sample units are visited in the field, it may be found that some are not in the expected strata. If these units are reclassified into the correct strata, then this means that all population units within the new stratum no longer have the same probability of being sampled. Hence, the sampling design is changed, with the possibility of introduction of some estimation bias. Another possible problem with stratified sampling is that after the data are collected, it is desired to do an analysis with some other form of stratification or some analysis that assumes that the data come from simple random sampling. This is always a possibility, and it led Overton and Stehman (1995) to argue strongly in favor of using simple sampling designs with limited or no stratification.

Generally, the types of stratification that should be considered are those based on spatial location, areas within which the population is expected to be fairly uniform, and the size of sampling units. For example, in sampling an animal population over a large area, it is natural to take a map and partition the area into a few apparently homogeneous strata based on factors such as altitude and vegetation type. In sampling insects on trees, it might make sense to stratify on the basis of small, medium, and large tree diameters. In sampling households, a town can be divided into regions within which the age and class characteristics are relatively uniform. Usually, the choice of how to stratify is just a question of common sense.

Assume that K strata have been chosen, with the ith of these having size N_i and the total population size being $\Sigma N_i = N$. Then, if a random sample with size n_i is taken from the ith stratum, the sample mean $\bar{y}_i$ will be an unbiased estimate of the true stratum mean μ_i with estimated variance $\widehat{\text{Var}}(\bar{y}_i) = (s_i^2 / n_i)$ $(1 - n_i/N_i)$ where s_i is the sample standard deviation of Y within the stratum. These results follow by simply applying the results discussed previously for simple random sampling to the ith stratum only.

In terms of the true strata means, the overall population mean is the weighted average

$$\mu = \sum_{i=1}^{K} N_i \mu_i / N, \tag{2.22}$$

and the corresponding sample estimate is

$$\bar{y}_s = \sum_{i=1}^{K} N_i \bar{y}_i / N, \tag{2.23}$$

with estimated variance

$$\widehat{\text{Var}}(\bar{y}_s) = \sum_{i=1}^{K} N_i^2 \, \widehat{\text{Var}}(\bar{y}_i) / N^2. \tag{2.24}$$

The estimated standard error of $\bar{y}_s$ is $\widehat{\text{SE}}(\bar{y}_s)$, the square root of the estimated variance, and an approximate $100(1 - \alpha)\%$ confidence interval for the population mean is given by

$$\bar{y}_s \pm z_{\alpha/2} \, \widehat{\text{SE}}(\bar{y}_s), \tag{2.25}$$

where $z_{\alpha/2}$ is the value exceeded with probability $\alpha/2$ for the standard normal distribution.

It sometimes happens that the population being dealt with is effectively infinite. For example, if samples are taken at point locations in a field, then the number of possible points can be regarded as infinite. In that case, Equations (2.22) to (2.24) can be modified by replacing N_i/N by the proportion of the population in stratum i. Hence, for sampling points in a field, this would be the proportion of the area in the whole field that is in stratum i. Also, in this situation the finite population corrections would not be needed because n_i/N_i would be zero for all strata.

If the population total is of interest, then this can be estimated by

$$\hat{T}_s = N\bar{y}_s \tag{2.26}$$

with estimated standard error

$$\widehat{\text{SE}}(\hat{T}) = N \, \widehat{\text{SE}}(\bar{y}_s). \tag{2.27}$$

Again, an approximate $100(1 - \alpha)\%$ confidence interval takes the form

$$\hat{T}_s \pm z_{\alpha/2} \, \widehat{\text{SE}}(\hat{T}). \tag{2.28}$$

If the sample sizes from the different strata are taken in proportion to the strata sizes, then this is called stratified sampling with proportional allocation. The samples are self-weighting in the sense that the estimates of the overall mean and the overall proportion are the same as what is obtained by lumping the results from all the strata together as a single sample. However, the simple random sampling variance formulas are not correct, and the stratified random sampling variance formulas should be used instead.

Although proportional allocation is often used because it is convenient, it is not necessarily the most efficient use of resources. One result applies if the total cost of a survey consists of a fixed cost F, and costs that are proportional to sample sizes in strata so that *Total Cost* $= F + \Sigma c_i n_i$, where c_i is the cost of sampling one unit from stratum i. Then, it can be shown that to either (1) achieve a given level of precision for estimating the overall population mean at the least cost or (2) to gain the maximum precision for a fixed total cost, the sample size in the ith stratum should be made proportional to $N_i \sigma_i / \sqrt{c_i}$. Use of this result requires approximate values for strata variances and knowledge of sampling costs. If these variances and costs are the same in all strata, then proportional allocation is optimal. For more details about how to apply the result, and optimum stratified sampling in general, see the work of Cochran (1977) or Scheaffer et al. (2011).

EXAMPLE 2.6 Stratified Sampling of Strip Transects

Consider again the situation discussed concerning the winter mortality of deer as estimated by counting dead deer in a sampled strip transect 1 km long and 60 m wide. Such a study might well include stratification based on habitat, on the assumption that dead deer are more likely to be found in some types of habitat than in others.

Suppose that this is the case and that a study area is divided into three habitat types. In type I habitat, there are $N_1 = 20{,}000$ potential strip transects, of which $n_1 = 20$ are randomly chosen for sampling. In type II habitat, there are $N_2 = 15{,}000$ strip transects, of which $n_2 = 15$ are randomly chosen for a sampling. Finally, in type III habitat there are $N_3 = 15{,}000$ strip transects, of which $n_3 = 15$ are randomly chosen for sampling. Suppose further that the number of dead deer found is as shown in Table 2.2. Note also the values of strata sample sizes n_i, sample means $\bar{y}_i$, sample standard deviations s_i, estimated variances of sample means $\mathrm{Var}(\bar{y}_i)$, and stratum sizes N_i, which are shown in the table.

Using Equation (2.23) and the information in Table 2.2, the estimated mean number of dead deer per transect for the whole population is $\bar{y}_s = (20{,}000/50{,}000) \times 0.90 + (15{,}000/50{,}000) \times 0.27 + (15{,}000/50{,}000) \times 0.33 = 0.540$, with estimated variance $\mathrm{Var}(\bar{y}_s) = (20{,}000/50{,}000)^2 \times 0.036 + (15{,}000/50{,}000)^2 \times 0.013 + (15{,}000/50000)^2 \times 0.025 = 0.00934$. The estimated standard error is then $\mathrm{SE}(\bar{y}_s) = \sqrt{0.00934} = 0.097$, and an approximate 95% confidence interval for the population mean number of deer per transect is $0.540 \pm 1.96 \times 0.097$ or 0.35 to 0.73. It follows that an estimate of the total number of dead deer in the entire study region is

TABLE 2.2

Results from a Strip Transect Survey of Winter Mortality of Deer

Transect	Habitat		
	I	II	III
1	1	1	0
2	1	0	0
3	2	0	0
4	2	0	0
5	0	1	0
6	0	0	0
7	0	0	0
8	0	0	0
9	1	1	0
10	2	0	1
11	2	1	0
12	0	0	2
13	2	0	1
14	1	0	0
15	0	0	1
16	2		
17	0		
18	0		
19	1		
20	1		
n	20	15	15
$\bar{y}_i$	0.90	0.27	0.33
s_i	0.85	0.46	0.62
$\text{Var}(\bar{y}_i)$	0.036	0.014	0.025
N	20,000	15,000	15,000

Note: The observations are the number of dead deer found in 1 km by 60 transects in three different habitat types.

$\bar{T}_y = 50{,}000 \times 0.540 = 27{,}000$, with estimated standard error $\widehat{\text{SE}}(\bar{T}_s) = 50{,}000 \times 0.097 = 4{,}850$, and an approximate 95% confidence interval for the population total is $27{,}000 \pm 1.96 \times 4{,}850$ or 17,500 to 36,500.

2.9.1 Poststratification

With some variables that are suitable for stratification, it is difficult to know the strata to which units belong until a survey has been conducted, although the strata sizes are known accurately for the population. For example, the stratification of sample quadrats on the basis of habitat type may not be possible until the quadrats have been visited.

One possible procedure in this case is to use poststratification, whereby a simple random sample of n is taken from the entire population, and the sampled units are classified into K strata. Then, the stratified sampling estimator $\bar{y}_s = \Sigma N_i \bar{y}_i / N$ is used to estimate the population mean.

Poststratification is almost as precise as stratified sampling with proportional allocation providing that the sample size is larger than about 20 in each of the strata. Furthermore, the stratified sampling equation for the variance of $\bar{y}_s$ is still approximately correct.

2.9.2 Stratified Sampling for Proportions

Stratified sampling can also be used with the estimation of proportions. If $\hat{p}_i$ is the sample proportion in stratum i, then this is an unbiased estimator of the stratum proportion p_i with estimated variance

$$\widehat{\mathrm{Var}}(p_i) = \{\hat{p}_i(1-\hat{p}_i)/n_i\}\{1 - n_i/N_i\}. \tag{2.29}$$

An unbiased estimator of the overall population proportion p is then

$$\hat{p} = \sum_{i=1}^{K} N_i \hat{p}_i / N, \tag{2.30}$$

with estimated variance

$$\widehat{\mathrm{Var}}(\hat{p}) = \sum_{i=1}^{K} N_i^2 \, \widehat{\mathrm{Var}}(\hat{p}_i) / N^2, \tag{2.31}$$

and estimated standard error $\widehat{\mathrm{SE}}(\hat{p}) = \sqrt{\widehat{\mathrm{Var}}(\hat{p})}$. An approximate $100(1 - \alpha)\%$ confidence interval for the true population proportion is given by

$$\hat{p} \pm \{z_{\alpha/2} \, \widehat{\mathrm{SE}}(\hat{p})\}, \tag{2.32}$$

where $z_{\alpha/2}$ is the value that is exceeded with probability $\alpha/2$ with the standard normal distribution.

In practice, if the cost to sample a unit is the same for all strata, then the gain from stratified random sampling over simple random sampling is small unless the proportions vary greatly with the strata. However, these equations might still be useful for surveys that include the estimations of means and proportions at the same time. Then, the value of stratification might be most obvious with the estimation of the mean values, but still the sample must be treated as stratified for estimating the required proportions.

2.10 Systematic Sampling

Systematic sampling can be carried out whenever a population can be listed in order or it covers a well-defined spatial area. In the former case, every kth item in the list can be sampled, starting at an item chosen at random from the first k. In the second case, sampling points can be set out in a systematic pattern to cover the area.

There are two reasons why systematic sampling is sometimes used in preference to random sampling. First, systematic sampling is often easier to carry out than random sampling. Second, it seems likely that a systematic sample will be more representative than a random sample, and hence more precise, because it gives uniform coverage of the whole population of interest.

As an example, consider the situation that is shown in Figure 2.2. The left-hand part of the figure shows the positions of 16 randomly selected plots in a rectangular study area. The middle part shows a stratified sample for which the study region is divided into four equal size strata, and four sample plots are placed randomly within each of these. The right-hand part of the figure shows a systematic sample for which the 16 sample locations are chosen over the study area with a fixed pattern. Clearly, stratified sampling has produced better control than random sampling in terms of the way that the sample points cover the region, but not as much control as systematic sampling.

Systematic sampling has the disadvantage of not allowing any simple determination of the level of sampling errors unless it is assumed that the items in the population are in a more or less random order. If that is the

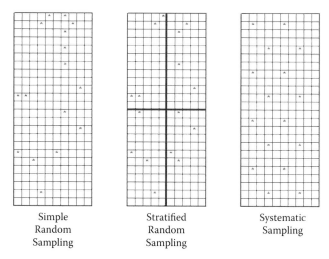

| Simple Random Sampling | Stratified Random Sampling | Systematic Sampling |

FIGURE 2.2
Comparison of simple random sampling, stratified random sampling, and systematic sampling for plots in a rectangular study region, with chosen plots indicated by an asterisk *.

case, then a systematic sample can be treated as effectively a simple random sample, and the various results given previously for this type of sampling can all be used.

Another possibility involves the estimation of a sampling variance by replicating a systematic sample. For example, suppose that a 10% sample of a population is required, where the population units are listed in some order. Rather than taking every 10th item, starting with one of the items 1 to 10, randomly chosen, it might be possible to take 20 systematic samples, each starting at a different randomly chosen item in the first 200, and sampling every 200th item from that point. The population can then be thought of as consisting of $N = 200$ clusters from which 20 are randomly sampled. Inferences concerning the population mean and total can be made in the usual manner for simple random sampling.

If systematic samples are analyzed as if they are simple random samples, then the sample variance $\Sigma(y_i - \bar{y})^2/(n - 1)$ will tend to be an overestimate of the true variance if the population shows a trend and might be an underestimate if the population has periodic effects that match or nearly match the distance between samples. However, if sample points are far enough apart to be more or less independent, then systematic and random sampling will be almost equivalent. In most situations, periodicity is unlikely to be a problem. It would require that, for some reason, the observations at the systematically determined sample locations tend to be higher or lower than observations in general.

One alternative to treating a systematic sample as effectively a random sample involves combining adjacent points into strata, as indicated in Figure 2.3, which is essentially just a poststratification of the data. The population mean and standard error are then estimated using the usual equations for stratified random sampling. The assumption being made is that the sample within each of the imposed strata is equivalent to a random sample.

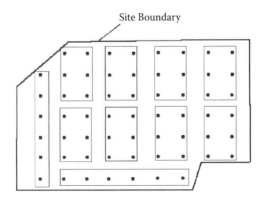

FIGURE 2.3

Grouping sample points from a systematic sample so that it can be analyzed as a stratified sample. The sample points • are grouped here into 10 strata, each containing six points.

It is most important that the strata are defined without taking any notice of the values of observations because otherwise bias will be introduced into the variance calculation. For example, if the strata are chosen to ensure that large observations only occur in some of the strata, then this will tend to lead to the within-strata variances being too small.

If the number of sample points or the area is not the same within each of the strata, then the estimated mean from the stratified sampling equations will differ from the simple mean of all of the observations. This is to be avoided because it will be an effect that is introduced by a more or less arbitrary system of stratification. The estimated variance of the mean from the stratified sampling equations will inevitably depend on the stratification used, and under some circumstances, it might be necessary to show that all reasonable stratifications give about the same result.

Another alternative to treating a systematic sample as a simple random sample involves joining the sample points with a serpentine line that joins neighboring points and passes only once through each point. This method was described by Manly (2009, Section 2.9) and usually seems to give results that are rather similar to the stratified sampling approach. Another possibility for the analysis of a systematic sample involves estimating the mean over an area using a geostatistical method, as reviewed by Manly (2009, Chapter 9). This will require some specialized computer software to perform the calculations.

EXAMPLE 2.7 Concentrations of Trichloroethylene in Groundwater

Kitanidis (1997, p. 15) gives the values of trichloroethylene (TCE) in groundwater samples in a fine-sand surficial aquifer. Here, 40 of these observations in a semisystematic rectangular grid are considered, as shown in Figure 2.4.

Treating the data as coming from a random sample of size 40, the sample mean is $\bar{y} = 5609.1$ ppb, with a sample standard deviation of $s = 12,628.9$ and an estimated standard error for the mean of $SE(\bar{y}) = s\sqrt{40} = 1996.8$. An approximate 95% confidence interval for the true mean concentration of TCE over the area is then $\bar{y} \pm 1.96.SE(\bar{y})$, which gives the range from 1695.3 to 9522.8. This range is wide because of the large amount of variation in the TCE values.

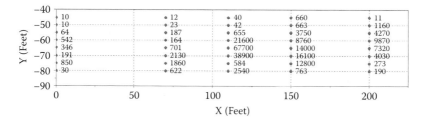

FIGURE 2.4
Concentrations of TCE (ppb, parts per billion) measured at 40 locations in a fine-sand surficial aquifer.

TABLE 2.3

Strata Means, Standard Deviations (SD), and Standard Errors (SE) for Means when the Observations for TCE Shown in Figure 2.4 Are Poststratified into Ten Strata Based on Treating Each of the Columns of Data in the Figure as Consisting of Two Strata of Four Observations Each

Stratum					Mean	SD	SE
1	10	10	64	542	156.5	258.3	129.1
2	346	191	850	30	354.3	354.8	177.4
3	12	23	187	164	96.5	91.8	45.9
4	701	2,130	1,860	622	1,328.3	778.4	389.2
5	40	42	655	21,600	5,584.3	10,681.1	5,340.5
6	67,700	38,900	584	2,540	27,431.0	32111.5	16,055.8
7	600	663	3,750	8,760	3,443.3	3,837.4	1,918.7
8	14,000	16,100	12,800	763	10,915.8	6,904.5	3,452.3
9	11	1,160	4,270	9,870	3,827.8	4,411.7	2,205.8
10	7,320	4,030	273	190	2,953.3	3,418.0	1,709.0
				Overall[a]	5,609.1		1,760.4

[a] The overall mean is the mean of the values for the individual strata because the strata sizes are all four and each has the same weight. The overall standard error is the estimated standard error of the overall mean based on the variance calculated using Equation (2.24), with $N_i/N = 1/10$.

An alternative analysis involves poststratification of the sample. For example, each of the columns of observations can be divided into two strata of four observations each based on the top four and the bottom four observations. There are then ten strata, with means, standard deviations, and standard errors of the means as shown in Table 2.3. The estimated mean TCE concentration is 5609.1, which is the same as obtained by treating the data as coming from a random sample. The estimated standard error of the mean is 1760.4, which is less than the value of 1996.8 that was obtained assuming a random sample. The approximate 95% confidence interval for the true mean based on the stratified sample is given by 5609.1 ± 1.96(1760.4), that is, 2158.7 to 9059.4 is therefore narrower than the interval calculated assuming a random sample.

In this example, the TCE concentrations are assumed to be effectively made at a point in the sample area. There is then an infinite number of possible sample points in each of the strata so that the standard errors shown in Table 2.3 are just $SD/\sqrt{4}$, without any finite population corrections.

2.11 Some Other Design Strategies

So far in this chapter, the sample designs considered are simple random sampling, stratified random sampling, and systematic sampling. There are also a number of other design strategies that are sometimes useful. Here, a few of these are briefly mentioned. For further details on these methods, see Manly (2009, Chapter 2). The following chapters also describe some other strategies.

With cluster sampling, groups of sample units that are close in some sense are randomly sampled together and then all measured. The idea is that this will reduce the cost of sampling each unit so that more units can be measured than would be possible if they were all sampled individually. This advantage is offset to some extent by the tendency of sample units that are close together to have similar measurements. Therefore, in general, a cluster sample of n units will give estimates that are less precise than a simple random sample of n units. Nevertheless, cluster sampling may give better value than the sampling of individual units in terms of what is obtained for a fixed total sampling effort.

With multistage sampling, the sample units are regarded as falling within a hierarchic structure. Random sampling is then conducted at the various levels within this structure. For example, suppose that there is interest in estimating the mean of some water quality variable in the lakes in a large area, such as a whole country. The country might then be divided into primary sampling units consisting of states or provinces, each primary unit might then consist of a number of counties, and each county might contain a certain number of lakes. A three-stage sample of lakes could then be obtained by first randomly selecting several primary sampling units, next randomly selecting one or more counties (second-stage units) within each sampled primary unit, and finally randomly selecting one or more lakes (third-stage units) from each sampled county. This type of sampling plan might be useful when a hierarchic structure already exists or when it is simply convenient to sample at two or more levels.

The technique of composite sampling is valuable when the cost of selecting and acquiring sample units is much less than the cost of analyzing them. What this involves is mixing several samples from approximately the same location and then analyzing the composite samples. For example, sets of four samples might be mixed so that the number of analyses is only one-quarter of the number of samples. This should have little effect on the estimated mean providing that samples are mixed sufficiently so that the observation from a composite sample is close to the mean for the samples that it contains. However, there is a loss of information about extreme values for individual sample units because of dilution effects. If there is a need to identify individual samples with extreme values, then methods are available to achieve this without the need to analyze every sample.

Ranked set sampling is another method that can be used to reduce the cost of analysis in surveys. The technique was originally developed for the estimation of the biomass of vegetation (McIntyre, 1952), but the potential uses are much wider. It relies on the existence of an inexpensive method of assessing the relative magnitude of a small set of observations to supplement expensive accurate measurements.

As an example, suppose that 90 uniformly spaced sample units are arranged in a rectangular grid over an intertidal study area, and that it is necessary to estimate the average barnacle density. A visual assessment is made of the density on the first three units, which are then on that basis ordered from the one with the lowest density to the one with the highest density. The density is then determined accurately for the highest ranked unit. The next three units are then visually ranked in the same way, and the density is then determined accurately for the unit with the middle of the three ranks. Next, sample units 7, 8, and 9 are ranked and the density determined accurately for the unit with the lowest rank. The process of visually ranking sets of three units and measuring first the highest-ranking unit, then the middle-ranking unit, and finally the lowest ranking unit is then repeated using units 10 to 18, units 19 to 27, and so on. After the completion of this procedure on all 90 units, a ranked set sample of size 30 is available based on the accurate estimation of density. This sample is not as good as would have been obtained by measuring all 90 units accurately, but it should have considerably better precision than a standard sample of size 30.

Ratio estimation is often used to allow for the varying sizes of sample units where the value of a variable of interest Y is assumed to be approximately proportional to the size of the sample unit that it is recorded in U, so that $Y \approx RU$ where R is a constant. For example, the number of animals might be counted in patches of vegetation of different sizes, with the assumption that the expected number of animals in a patch is proportional to the area of the patch. Another possibility with ratio estimation is that the size of an animal or plant population needs to be estimated, and that it is expensive to accurately determine the abundance in a sample unit Y. However, there is some other variable U that is not expensive to measure that is approximately proportional to the abundance. For example, this second variable might just be an estimate based on a quick visual survey of the sample unit. The technique used involves making the visual survey of either all or a large sample of the population of sample units and making the accurate determination of abundance on a small number of these units. The idea then is to adjust the estimated abundance based on the small sample using the relationship between the accurate measurements of abundance and the values of the inexpensive-to-measure variable. This is done by first estimating the ratio $\hat{R} = \bar{y} / \bar{u}$ using the sample units for which both Y and U are known and then multiplying this by the mean value of U based on either all units in the population or the mean of a large sample of U values to give the estimated mean of Y as $\bar{y}_{ratio} = \hat{R}U_{mean}$.

Ratio estimation assumes that the ratio of the variable of interest Y to the subsidiary variable U is approximately constant for the sample units in the population. A less-restrictive assumption is that Y and U are approximately related by a linear equation of the form $Y = \alpha + \beta U$, in which case regression estimation can be used. As for ratio estimation, a sample of units is selected and values of Y and U obtained. Ordinary linear regression is then used to estimate the equation $Y = a + bU$. The mean of U, μ_u, is estimated either from a much larger sample of units or for all units in the population, and the regression estimator of the mean of Y is $\bar{y}_{reg} = \bar{y} + b(\mu_u - \bar{u})$ where $\bar{y}$ and $\bar{u}$ are the means for the smaller sample of Y and U values. This can be interpreted as $\bar{y}$ from the small sample corrected by $b(\mu_u - \bar{u})$ to allow for a low or high mean of the U values in the small sample.

2.12 Unequal Probability Sampling

Situations do arise for which the nature of the sampling mechanism makes random or systematic sampling impossible because the availability of sample units is not under the control of the investigator. In particular, cases occur for which the probability of a unit being sampled is a function of the characteristics of that unit, which is called unequal probability sampling. For example, large units might be more conspicuous than small ones, so that the probability of a unit being selected depends on its size. If the probability of selection is proportional to the size of units, then this special case is called size-biased sampling.

It is possible to estimate population parameters allowing for unequal probability sampling. Thus, suppose that the population being sampled contains N units with values $y_1, y_2, \ldots y_N$ for a variable Y, and that sampling is carried out so that the probability of including y_i in the sample is p_i. Assume that estimation of the population size N, the population mean μ_y, and the population total T_y is of interest, and that the sampling process yields n observed units. Then, the population size can be estimated by

$$\hat{N} = \sum_{i=1}^{n} (1/p_i), \tag{2.33}$$

the total of Y can be estimated by

$$\hat{T}_y = \sum_{i=1}^{n} (y_i / p_i), \tag{2.34}$$

and the mean of Y can be estimated by

$$\hat{\mu}_y = \hat{T}_y / \hat{N} = \sum_{i=1}^{n} (y_i / p_i) / \sum_{i=1}^{n} (1/p_i). \tag{2.35}$$

Also, if the probability of selection is proportional to a variable X that has the known value of x_i for the ith sampled unit, then Equation (2.35) becomes

$$\hat{\mu}_y = \hat{T}_y / \hat{N} = \sum_{i=1}^{n} (y_i / p_i) / \sum_{i=1}^{n} (1 / p_i), \tag{2.36}$$

and the estimated mean of the variable X itself becomes

$$\hat{\mu}_x = n / \sum_{i=1}^{n} (1 / x_i). \tag{2.37}$$

These estimators are called Horvitz–Thompson estimators (Horvitz and Thompson, 1952). They provide unbiased estimates of the population parameters because of the weight given to different observations. For example, suppose that there are a number of population units with $p_i = 0.1$. Then, it is expected that only 1 in 10 of these units will appear in the sample of observed units. Consequently, the observation for any of these units should be weighted by $1/p_i = 10$ to account for those units that are missed from the sample. Variance equations for all three estimators were provided by McDonald and Manly (1989), who suggested that replications of the sampling procedure will be a more reliable way of determining variances. The book by Thompson (2012) gives a comprehensive guide to the many situations that occur for which unequal probability sampling is involved.

3

Adaptive Sampling Methods

Jennifer Brown

3.1 Introduction

Environmental managers' need for accurate information about ecological populations is increasing as the demand for cost-effective management expands. The ability to detect the current status of a population is essential for managers to be able to decide when to take appropriate management action and for assessing the success—or otherwise—of alternative management interventions. Information is needed about population status for scientists to assess and evaluate alternative population models and for understanding population dynamics. In the previous chapter, the idea of sampling and selecting a portion of the population compared with assessing and measuring every individual in the population was introduced. Here, we introduce a range of sampling designs that are particularly relevant to ecological sampling and in fact have been designed mostly with these situations in mind (Thompson, 2003). To be more precise, these designs were developed for surveying populations that are rare and clustered.

Ecologists' interest in rare and clustered populations is because these populations often represent the extremes of biological processes, such as the early incursion of invasive species or the near extinction of endangered species. They can also be the most challenging to survey because without some targeted field effort, most of the sample units, by definition, will not contain the rare species of interest. Survey designs that assist in targeting field effort to where the species of interest is should improve the quality and quantity of information from the survey.

There is no one definition for rare and clustered populations that all statisticians and ecologists agree on, but generally, these are populations that are difficult to detect (in fact, they may not be rare but sightings of them are rare) or populations that are sparse in some sense. Sparseness can come about because there are few individuals in the population or because the population covers a large area (McDonald, 2004).

Adaptive sampling designs involve the survey team changing or adapting the procedure used to select sample units as information comes to hand during the course of sampling. The key point here is that the protocol of what to sample and where may change as the survey evolves, and the field crew needs to be able to adapt to this change. The statistical properties of all the designs discussed are well documented, and they all fall within the realm of probability-based sampling.

There are two categories of adaptive sampling: adaptive searching and adaptive allocation (Salehi and Brown, 2010). Adaptive searching includes designs for which the neighborhood of certain sample units is of special interest (e.g., a rare plant is found in the sample unit and the search shifts to focus on the surrounding units); adaptive allocation involves allocation of additional survey effort to the general area of interest, such as a stratum that is thought to contain rare plants.

Interest in adaptive sampling peaked with the development of adaptive cluster sampling (Thompson, 1990), although adaptive designs were suggested well before this (e.g., Francis, 1984). I begin by discussing adaptive cluster sampling and then move into some of the other adaptive allocation designs to illustrate the range and versatility of adaptive sampling.

3.2 Adaptive Cluster Sampling

Adaptive cluster sampling was first introduced by Thompson (1990). The design was developed for sampling populations that are rare and clustered. It is similar to cluster sampling, in which a cluster of units is selected, and either the entire cluster or a portion of it is sampled. The typical example used to explain cluster sampling is surveying children in a school. Classes are natural clusters of children, so a selection of classes is chosen, and children within classes are surveyed. In adaptive cluster sampling, clusters are selected, but the difference is that the size, location, and total number of clusters are not known.

In its simplest form, adaptive cluster sampling starts with a random sample. Prior to sampling, a threshold value C is chosen, and if any of the units in the initial sample meet or exceed this threshold, $y_i \geq C$, then neighboring units are sampled. Any unit that meets or exceeds this threshold is considered to have "met the condition." If any of these neighboring units meets this condition, their neighboring units are selected and so on. In this way, as sampling continues for any cluster that is detected in the initial sample, the edge of the cluster is delineated, and the cluster itself will be sampled. The final sample is the collection of clusters that were detected in the initial sample, along with any of the sample units that were in the initial sample but below the threshold. This design has considerable intuitive appeal for clustered populations. Once the rare plant or animal is observed in the initial search, survey effort is then focused on the immediate area surrounding the occupied sample unit where the rare plant or animal is most likely to be found. The statistical properties of the process for selecting the initial sample unit and the neighborhood search mean that the final sample will be unbiased.

The term *cluster* in adaptive cluster sampling is used to refer to the collection of contiguous sample units (called a network) and the surrounding edge units. A network is the set of sample units that surround the unit in the initial sample that triggered neighborhood searching. All the network units will have met the condition. The neighborhood units that are searched but have values below the threshold are the edge units. Calculating the sample estimators requires the definition that any unit in the initial sample that does not meet the condition is also considered a network and it has a size of one.

The study area can be divided into distinct networks that do not overlap. Some of these networks will be only one unit in size; others will be larger than one unit. In Figure 3.1, with the condition $y_i \geq 1$, the 200-quadrat study area can be divided into 187 distinct networks. The top right-hand corner has a network of size 5, the aggregate to the left of this is a network of size 7, and there is a network of size 4. All the other quadrats, even those with counts in them, are considered networks of size 1.

In simple random sampling, the probability of selecting any sample unit is the same for each unit. With adaptive cluster sampling, the probability of a unit being selected is more complicated because there can be many different ways for a unit to appear in the sample. The sample unit could be in the sample because it was in the initial selection or because any one of its neighboring units was in the initial sample. The survey design is a form of unequal probability sampling because not all sample units have the same chance of being selected. In Figure 3.1, the large size 5 network in the top right-hand corner will appear in the final sample if any of the five units are in the initial sample, whereas a size 1 network has only one chance of being in the final sample.

There are two estimators that can be used, the Horvitz–Thompson estimator and the Hansen–Hurwitz estimator, although the Horvitz–Thompson estimator is preferred (Thompson and Seber, 1996). The Horvitz–Thompson estimate of the population total is calculated using the equation

$$\hat{\tau} = \sum_{k=1}^{K} \frac{y_k^* z_k}{\alpha_k},$$ (3.1)

where y_k^* is the total of the y values in network k; z_k is an indicator variable equal to one if any unit in the kth network is in the initial sample and zero otherwise; and α_k is the initial intersection probability. This initial intersection probability is the probability that at least one of the units in the network

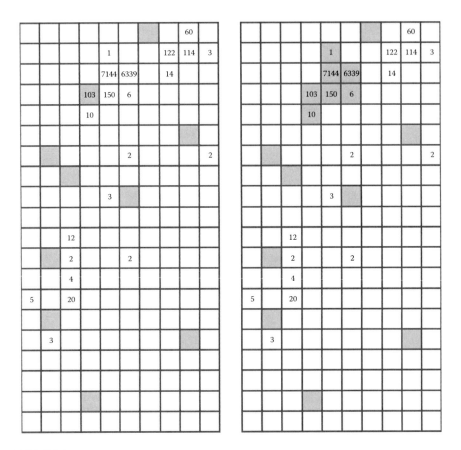

FIGURE 3.1
A population of blue-winged teal ducks. (From Smith, D.R., Conroy, M.J., and Brakhage, D.H. *Biometrics* 51: 777–788, 1995.) There are two hundred 25-km² quadrats with the counts of observed ducks shown. The left side shows an initial sample of 10 quadrats, and the right side shows the final sample. One network larger than one quadrat in size is selected in the final sample (the network has seven quadrats in it). The condition to trigger adaptive selection was $y_i \geq 1$. A neighborhood was defined as the surrounding four quadrats.

will be in the initial sample. For the kth network of size x_k, the initial intersection probability is

$$\alpha_k = 1 - \binom{N - x_k}{n} \bigg/ \binom{N}{n},\tag{3.2}$$

where N is the size of the study area, x_k is the size of the network, and n is the size of the initial sample. The estimate of the variance of the estimated population total involves the joint inclusion probabilities α_{jk} and calculating the probability that both network j and k appear in the initial sample where

$$\alpha_{jk} = 1 - \left[\binom{N - x_j}{n} + \binom{N - x_k}{n} - \binom{N - x_j - x_k}{n} \right] / \binom{N}{n}, \tag{3.3}$$

where $j \neq k$ and $\alpha_{jk} = \alpha_j$, when $j = k$. The estimator of the variance is

$$\widehat{\text{Var}(\hat{\tau})} = \sum_{j=1}^{K} \sum_{k=1}^{K} y_j^* y_K^* \left(\frac{\alpha_{jk} - \alpha_j \alpha_k}{\alpha_j \alpha_k} \right) z_j z_k \tag{3.4}$$

In Figure 3.1, which uses blue-winged teal as an example (Brown, 2011; Smith et al., 1995), the initial sample is 10 quadrats, $n = 10$; the threshold condition is $y_i \geq 1$; and the definition of the neighborhood is the four surrounding quadrats. Only one quadrat in the initial sample triggered adaptive selection of the surrounding quadrats.

The final sample size is 16. Note that many more than 16 quadrats are assessed because the 4 neighboring quadrats around any occupied quadrat are checked. Only quadrats in the initial sample or in a selected network are used in calculating the sample estimators, and the other quadrats are the edge units. With a simple condition like $y_i \geq 1$, the quadrats only need to be checked to see if ducks are present or absent, but with a condition like $y_i \geq 10$, ducks within the units would need to be counted to know whether there were less than 10, something that may be more time consuming than simply checking for presence.

The Horvitz–Thompson estimate using Equation (3.1) of the population total from the sample in Figure 3.1 is

$$\hat{\tau} = \sum_{k=1}^{187} \frac{y_k^* z_k}{\alpha_k}$$

$$= \frac{13753 \cdot 1}{0.3056} + 0 + \ldots + 0$$

$$= 45003.$$

The only nonzero term in this equation is for the network of size 7. The values of all the other terms are zero. The other nine networks that were selected in the initial sample were only one quadrat in size and $y^* = 0$. All the other networks that were not selected have $z_k = 0$.

The initial intersection probability, Equation (3.2), for the size 7 network is calculated as

$$\alpha_1 = 1 - \binom{200 - 7}{10} / \binom{200}{10} = 0.3056.$$

The estimator of the variance for the duck population is fairly straightforward because there is only one network sampled where the y^* is not zero. The joint inclusion probability using Equation (3.3) for the network of size 7 with itself is

$$\alpha_{11} = \alpha_1 = 0.3056.$$

The joint inclusion probabilities for networks that were not selected do not need to be calculated because for these, $z_k = 0$. The estimator of the variance from Equation (3.4) is therefore

$$\widehat{Var}(\hat{\tau}) = \sum_{j=1}^{187} \sum_{k=1}^{187} y_j^* y_k^* \left(\frac{\alpha_{jk} - \alpha_j \alpha_k}{\alpha_j \alpha_k} \right) z_j z_k$$

$$= y_1^* y_1^* \left(\frac{\alpha_{11} - \alpha_1 \alpha_1}{\alpha_1 \alpha_1} \right) z_1 z_1 + y_1^* y_2^* \left(\frac{\alpha_{12} - \alpha_1 \alpha_2}{\alpha_1 \alpha_2} \right) z_1 z_2 +$$

$$\ldots + y_{187}^* y_{187}^* \left(\frac{\alpha_{187\,187} - \alpha_{187} \alpha_{187}}{\alpha_{187} \alpha_{187}} \right) z_{187} z_{187}$$

$$= y_1^* y_1^* \left(\frac{\alpha_{11} - \alpha_1 \alpha_1}{\alpha_1 \alpha_1} \right) z_1 z_1 + y_1^* \cdot 0 \cdot \left(\frac{\alpha_{12} - \alpha_1 \alpha_2}{\alpha_1 \alpha_2} \right) z_1 z_2 +$$

$$\ldots + y_{187}^* y_{187}^* \left(\frac{\alpha_{187\,187} - \alpha_{187} \alpha_{187}}{\alpha_{187} \alpha_{187}} \right) \cdot 0$$

$$= 13753^2 \left(\frac{0.3056 - 0.3056^2}{0.3056^2} \right) \cdot 1 \cdot 1$$

$$= 4.2978 \cdot 10^8$$

There are various software packages that can be used for these calculations (e.g., Morrison et al., 2008).

A considerable amount of literature has been written on how to design an adaptive cluster sampling method to improve survey efficiency. How efficient the design will be depends on how clustered the population is; the more clustered the population is, the more efficient adaptive cluster sampling is compared with simple random sampling (Smith et al., 2004, 2011; Brown, 2003). Efficiency also depends on the survey design, including features such as the condition used to trigger adaptive selection; the definition of the neighborhood in which adaptive searching takes place (e.g., the surrounding two, four, or eight neighboring units); the size of the initial sample; and the size of the sample units. As a general rule, efficient designs will have the final sample size not excessively larger than the initial sample size and will have small networks. This can be achieved using large criteria for adapting and small neighborhood definitions (Brown, 2003). Recent reviews of adaptive cluster sampling were given by Smith et al. (2004) and Turk and Borkowski (2005).

So far, the discussion has been about adaptive cluster sampling for which the initial survey is a simple random sample. Adaptive cluster sampling can also be used with systematic sampling (Thompson, 1991a), stratified sampling (Brown, 1999; Thompson, 1991b) and two-stage sampling (Salehi and Seber, 1997). For example, in Figure 3.2 the blue-winged teal population is sampled using stratified adaptive cluster sampling. The site is delineated into two strata, and the initial sample of 10 quadrats allocated proportional to the relative stratum size. In the figure, the initial sample size of the top stratum is four quadrats, and the stratum below has an initial sample size of 6 quadrats. The final sample size in the lower stratum stays the same, and in the top stratum adaptive selection of additional quadrats resulted in a within-stratum final sample size of 10 quadrats.

One final point on the design of adaptive cluster sampling is a concern often raised with adaptive designs, namely, that the size of the final sample is not known prior to sampling, and this makes planning the fieldwork difficult.

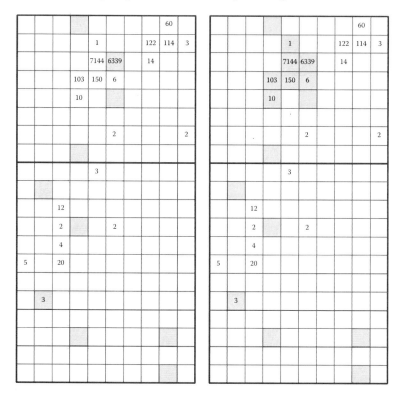

FIGURE 3.2
The population of blue-winged teal ducks (see Figure 3.1) sampled with stratified adaptive cluster sampling. The left side shows an initial sample of 10 quadrats with 4 in the top stratum and 6 in the lower stratum. The right side shows the final sample. One network larger than one quadrat in size is selected in the final sample (the network has seven quadrats in it). The condition to trigger adaptive selection was $y_i \geq 1$. A neighborhood was defined as the surrounding four quadrats.

Restricting the final sample size by a stopping rule has been discussed by Salehi and Seber (2002), Brown and Manly (1998), Lo et al. (1997), and Su and Quinn (2003). Another approach is an inverse sampling design, in which surveying stops once a set number of nonzero units has been selected (Seber and Salehi, 2004; Christman and Lan, 2001). More recently, Gattone and Di Battista (2011) have suggested a data-driven stopping rule that can be used when there is limited information about the rarity and clustering of the population.

There have been many applications of adaptive cluster sampling to a range of environmental situations. Each offers an insight into the design and application of adaptive cluster sampling. Some recent examples are the use of adaptive cluster sampling for surveys of freshwater mussels (Smith et al., 2003, 2011; Hornbach et al., 2010; Outeiro et al., 2008); rockfish (Hanselman et al., 2003); rare fish (Davis and Smith, 2011); a highly heterogeneous population of yellow perch (Yu et al., 2012); fish eggs (Smith et al., 2004; Lo et al., 1997); larval sea lampreys (Sullivan et al., 2008); larval walleye pollock (Mier and Picquelle, 2008); subtidal macroalgae (Goldberg et al., 2007); sediment load in rivers (Arabkhedri et al., 2010); forests (Talvitie et al., 2006; Acharya et al., 2000); deforestation rates (Magnussen et al., 2005); shrubs (Barrios et al., 2011); understory plants (Abrahamson et al., 2011); herbaceous plants (Philippi, 2005); annual plants (Morrison et al., 2008); bark beetle infestations (Coggins et al., 2011); giant panda habitat use (Bearer et al., 2008); marsupials (Smith et al., 2004); waterfowl (Smith et al., 1995); and herpetofauna (Noon et al., 2006); and in hydroacoustic surveys (Conners and Schwager, 2002).

3.3 Other Adaptive Sampling Designs

Adaptive sampling can be applied to stratified and two-stage sampling in the same way that the adaptive component is added to simple random sampling in adaptive cluster sampling. In stratified and two-stage sampling, the population is partitioned into distinct subareas, called strata in stratified sampling and primary units in two-phase sampling. The two designs, stratified sampling and two-phase sampling, are similar in how the survey is conducted in this respect. With stratified sampling, all of the strata are surveyed; in two-phase sampling, only some of the primary units are surveyed.

The idea behind adaptive allocation sampling is that strata or primary units that contain clusters of the rare population are preferentially allocated extra survey effort. Ideally, if the locations and sizes of all the clusters were known, strata or primary unit boundaries could be marked around each cluster. Additional effort would then be allocated to the strata or primary units that contained clusters. More often, the locations and sizes of clusters are not known, and the study area is partitioned into strata and primary units in a way that minimizes the within-primary unit or the within-stratum variance.

The study area may be partitioned according to habitat features (e.g., so that the open grassed areas are in one stratum and the shrubs in another). Other considerations may be related to field logistics and natural features in the field, such as catchment boundaries or fence lines, that can be used. The size of primary units may represent what sampling can be achieved by the field crew in one day, simplifying planning fieldwork to a primary unit per day.

One of the early applications of adaptive sampling to a stratified design is the two-phase stratified design proposed by Francis (1984). In the initial phase, the survey area is partitioned into strata, and an initial survey is conducted in which effort is allocated among the strata according to an estimate of within-stratum variability as in conventional stratified sampling. The initial phase sample results are used to estimate within-stratum variance. The remaining sample units are added one by one to an individual stratum. At each step of this sequential allocation of sample units, the stratum that is allocated the unit is chosen on the basis of where the greatest reduction in variance will be.

The design is adaptive in that the preliminary information is used to update the estimate of the within-strata variability, and the remaining survey effort is allocated to the strata that will be most effective in reducing the overall sample variance. The adaptive allocation in the second phase is done to adjust or to make up for any shortcomings in the initial allocation of effort. For some populations, rather than using the within-stratum variance of the criteria for adaptive allocation, the square of the stratum mean is preferred (Francis, 1984). See also the work of Jolly and Hampton (1990) for a discussion of a similar design.

The estimates of the population total and the sample variance are derived from all the information collected in the two phases. There can be a small bias in the sample estimate, but this can be corrected by bootstrapping (Manly, 2004). Adaptive two-phase sampling has been compared favorably with adaptive cluster sampling (Yu et al., 2012; Brown, 1999). The design can also be used for surveying multiple species populations (Manly et al., 2002).

A related design was proposed by Smith and Lundy (2006) for a stratified sample of sea scallops, with the within-stratum mean from the first phase used to allocate a fixed amount of effort to strata where the mean was above a threshold value. They used the Rao–Blackwell method (Thompson and Seber, 1996) to derive an unbiased estimate for the population. Harbitz et al. (2009) used a similar design for surveys of Norwegian spring-spawning herring, for which extra survey transects were added to strata that had fish densities (estimated from acoustic data) over a threshold limit. Another recent design suggested for fishery surveys uses an optimization approach to minimize the average distance between sample points and then allocates second-phase effort according to a predefined abundance threshold (Liu et al., 2011).

Adaptive allocation has been used with two-stage sampling in a design called adaptive two-stage sequential sampling (Brown et al., 2008; Moradi and Salehi, 2010). Note that here two-stage sampling is used, whereas the previous discussion was about two-phase sampling. The two concepts are close. As in the two-phase stratified designs, an initial sample is taken from selected primary units. Then, in the second phase, additional units are allocated to the primary units in proportion to the number of observed units in that primary unit that exceed a threshold value $g_i\lambda$, where g_i is the number of sampled units in the ith primary unit that exceed the threshold value and λ is a multiplier. In two-stage sequential sampling (Salehi and Smith, 2005), the number of additional units allocated to the primary units is set at some fixed value. The predefined threshold that triggers adaptive allocation can be the value of the observed unit y_i or some auxiliary information related to y_i (Panahbehagh et al., 2011). For surveying a rare and clustered population, these designs allow the survey effort to be intensified at the locations where the population is found. The results of comparative simulation surveys showed gains in survey efficiency (measured by a reduced sample variance) when the adaptive component was added to the conventional two-stage design (Brown et al., 2008; Salehi et al., 2010; Smith et al., 2011).

In complete allocation stratified sampling, adaptive sampling is applied to a conventional stratified or two-stage design (Salehi and Brown, 2010). Starting with a conventional stratified design, if any unit in a stratum has a value that exceeds a threshold, the stratum is completely surveyed. This design simplifies adaptive sampling in two ways: First, the rule to decide whether a stratum is to be allocated additional survey effort does not require the first-phase survey in the stratum to be completed. Second, the instructions to the field crew about how much additional effort is required are simply to survey the entire stratum.

The complete allocation stratified design uses the best features of some of the previous adaptive designs. Adaptive cluster sampling has considerable intuitive appeal to a field biologist in surveys of rare and clustered populations because once a rare plant or animal is found, the neighboring area is sampled. In complete allocation, the neighboring area is completely searched, and what is considered the neighboring area is defined prior to sampling and constrained by the stratum boundary. In adaptive cluster sampling, the neighborhood is not defined prior to sampling and, for some populations, can be excessively large (Brown, 2003).

The estimate of the population total for complete allocation stratified sampling is

$$\hat{\tau} = \sum_{h=1}^{\gamma} \frac{y_h^*}{\pi_h} \tag{3.5}$$

where y_h^* is the total of the y values in the hth stratum, γ is the number of strata that were completely surveyed, and π_h is the probability the whole stratum h is selected, which is

$$\pi_h = 1 - \left(\begin{array}{c} N_h - m_h \\ n_h \end{array} \right) \Big/ \left(\begin{array}{c} N_h \\ n_h \end{array} \right). \tag{3.6}$$

For the hth stratum in Equation (3.6), N_h is the size of the stratum, m_h is the number of nonempty units in the stratum, and n_h is the size of the initial sample in the stratum; an unbiased variance estimator is

$$\widehat{\text{Var}}[\hat{\tau}] = \sum_{h=1}^{\gamma} \frac{(1-\pi_h)y_h^{*2}}{\pi_h^{2}}. \tag{3.7}$$

Complete allocation stratified sampling compares well with nonadaptive designs such as conventional stratified sampling and two-phase sampling and realizes considerable improvements in sample efficiency (Brown et al., 2012).

The example provided by Brown (2011), shown in Figure 3.3, considers a rare buttercup found in the South Island of New Zealand. The study area was divided into 12 strata with 25 units in each. A simple random sample

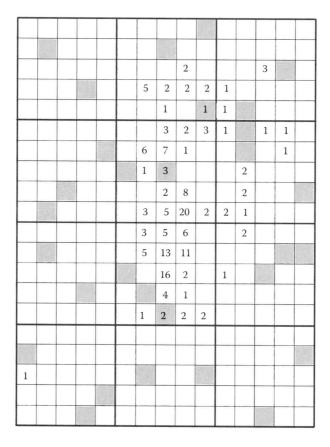

FIGURE 3.3
A population of Castle Hill buttercups. There are three hundred 100-m² quadrats with the counts of buttercups shown. The study area is sectioned into 12 strata, and three quadrats are sampled from each stratum.

of size 3 was taken from each stratum in phase one. From this first-phase information, three of the strata had sample units that contained buttercups. In the second phase, these three strata were surveyed completely, and the total final sample size was therefore $(3 \times 25) + (9 \times 3) = 102$.

For the three strata that did trigger adaptive sampling, π_h, the probability that whole stratum is selected (Equation 3.6) is:

$$\pi_5 = 1 - \frac{\binom{18}{3}}{\binom{25}{3}} = 0.645 \, ,$$

$$\pi_6 = 1 - \frac{\binom{11}{3}}{\binom{25}{3}} = 0.928, \text{ and}$$

$$\pi_7 = 1 - \frac{\binom{11}{3}}{\binom{25}{3}} = 0.928.$$

In the other nine strata, no plants were observed. The estimate of the total number of plants using Equation (3.5) is:

$$\hat{\tau} = \sum_{h=1}^{3} \frac{y_h^*}{\pi_h}$$

$$= \frac{15}{0.645} + \frac{66}{0.928} + \frac{73}{0.928}$$

$$= 172.99 \qquad .$$

The estimated variance of this estimate from Equation (3.7) is:

$$\widehat{\mathrm{Var}}[\hat{\tau}] = \sum_{h=1}^{3} \frac{(1 - \pi_h) y_h^{*2}}{\pi_h^2}$$

$$= \frac{(1 - 0.645)15^2}{0.645^2} + \frac{(1 - 0.928)66^2}{0.928^2} + \frac{(1 - 0.928)73^2}{0.928^2}$$

$$= 998.08 \qquad .$$

3.4 Discussion

Adaptive sampling is a broad term that encompasses a wide range of designs. The essential feature of the designs for rare and clustered populations that are discussed in this chapter is an improvement in sample efficiency. Adaptive sampling allows survey effort to be targeted to where any plant or animal of interest has been found, and this targeting leads to improvements in efficiency.

A number of different designs are introduced in this chapter, and the focus has been on how adaptive selection can be added to existing well-known designs. Adaptive cluster sampling is a technique leading from simple random sampling, stratified, systematic, or two-stage sampling, where additional survey effort is allocated to the immediate neighborhood of the sample units within which an individual is found (or where some measure from the sample unit exceeds a threshold value). The surveyor prespecifies the pattern of the neighborhood that is searched and the threshold value that triggers adaptive allocation. These survey protocol features have an important influence on the final sample size and sample efficiency. Various modifications to Thompson's (1990) original design have been suggested to constrain the final sample size.

Other adaptive designs discussed include designs for allocating effort in stratified and two-stage sampling. In two-phase stratified sampling, additional survey effort is allocated to certain strata on the basis of first-phase within-strata estimates. In adaptive two-stage sequential sampling, additional survey effort is allocated to primary units based on whether the primary unit estimate exceeds a threshold value.

Complete allocation stratified sampling uses the features of stratified sampling, where the population is categorized into strata. In the first phase, if any individuals are found within a stratum, the entire stratum is searched. Matching the size and shape of the strata as closely as possible to the size and shape of the likely clusters in the species of interest will ensure that the final survey effort is well targeted to searching where the species occurs. Even without this perfect match between the strata and species clusters, the survey method is still an efficient design.

There are many other adaptive designs than those discussed here. Other adaptive sampling designs can be found in the literature, often initiated by a very practical and real sampling problem. For example, Samalens et al. (2007) developed a sample plan for detecting beetle bark infestation. Additional survey effort was deployed along the edge of forest stands that were near piles of logs where the beetles were breeding. Yang et al. (2011) designed a forest survey based on adaptive cluster sampling for which, instead of searching the neighborhood when a sample unit (a forest plot) met the predefined condition, the size of the plot was increased. Adaptive line transect sampling was suggested for line transects to increase effort by following a zigzag pattern in sections of the line where a threshold abundance had been met (Pollard et al., 2002). They discussed surveys of harbor porpoises and how having a zigzag survey path that did not cross itself, that had no gaps, and was easily followed was an important consideration for shipboard surveys.

As interest in adaptive sampling continues to grow, the number and complexity of different designs will grow. To assist this expanding field of designs, Salehi and Brown (2010) suggested the use of the terms *adaptive searching* and *adaptive allocation* to distinguish two categories. Adaptive searching refers to designs such as adaptive cluster sampling, by which the neighborhood is searched. In contrast, in adaptive allocation, extra effort is initiated once a collection of units has been sampled (e.g., the stratum or the primary unit). The distinction between the two classes is based on where and when the decision to allocate extra effort can be made: immediately after an individual sample unit is measured or once a collection of units has been completely sampled. Similarly, the distinction between two-stage and two-phase sampling can be confusing. In this chapter, in two-stage sampling a selection of primary units is sampled, and in the second stage, a sample is taken within each. In two-phase sampling as discussed in this chapter, an initial sample is taken, and on the basis of information from that sample, additional effort is allocated in the second phase. In adaptive searching, the decision to conduct the second phase of sampling occurs concurrently with the first phase. In adaptive allocation, the decision to conduct the second-phase sampling occurs after the first phase.

All the designs discussed can be efficient, giving estimates of populations that have lower variance than the conventional design without adaptive selection. However, as with conventional sampling, the survey must be designed carefully to realize these gains in efficiency. Practical considerations will limit how complex the field protocol can be, and as any field biologist knows, there are advantages in keeping things simple. The final design chosen for a particular survey will be a balance between theoretical efficiencies and practical realities.

4

Line Transect Sampling

Jorge Navarro and Raúl Díaz-Gamboa

4.1 Introduction

Line transect sampling is intended not only for the estimation of the abundance per unit area of rare, mobile, difficult-to-detect animals but also is of value for the study of rare, difficult-to-detect plants, intertidal organisms, and so on (Burnham et al., 1980). The technique is related to variable circular plot sampling and is sometimes called distance sampling (Buckland et al., 2001). With line transect sampling, the basic idea is that an observer moves along a line through a study area, looking to the left and right for the animal or plant of interest. Line transects are walked, flown, or otherwise traversed, and the perpendicular distances to all detected items of interest are recorded. Combining these data with the assumption that all items on the line are detected, it is possible to correct the estimates of abundance per unit area for the items not detected. When an individual of the species of interest is detected, it is recorded, usually together with its distance from the line, because it is assumed that individuals that are far from the line are harder to detect than those that are close. This is one of the specialized ways that ecologists can use to estimate the density or the total number of animals or plants in a study area when it is not possible to simply count all the individuals and the standard sampling methods considered in Chapter 2 are for some reason not practical. For example, if the study area is very large and the species of interest is rare, then a random sample of quadrats in the study area may contain no individuals. However, if a long line is traveled through the area, then some individuals might be detected.

The size of items or other variables might influence the probability of detection, and such variables are sometimes included in the model that is used to correct for the items missed. The accurate estimation of the density or number of animals is always difficult using specialized methods like line transect sampling, and with all such methods, the final estimate obtained depends on the assumptions made. If these assumptions are not correct, then an estimate might be quite different from the true value.

4.2 Basic Procedures in Line Transect Sampling

Two types of data are recorded in line transect sampling, as shown in Figure 4.1. These are either (1) the perpendicular distances from the transect line x or (2) the sighting distances r and angles θ. However, studies based on sighting distances and angles have been found to be subject to biases and are only discussed briefly here.

The usual assumptions made with line transect sampling are the following:

1. All objects on the transect line are detected.
2. Objects do not move in response to the observer before the detection is recorded.
3. Objects are only counted once.
4. Objects are recorded at the point of initial detection.
5. Distances are measured without errors.
6. Transect lines are randomly located in the study area.

A further assumption sometimes made for the estimation of standard errors is that

7. Sightings are independent events, and the number of objects detected follows a Poisson distribution.

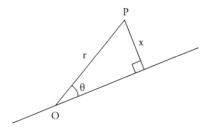

FIGURE 4.1
Graphical representation of statistics collected in line transect sampling. The letters O and P represent the positions of the observer and detected object, respectively. The perpendicular distance from the point P to the transect line is labeled x, the sighting distance from the observer to the point is labeled r, and the sighting angle is denoted by θ.

4.3 The Detection Function

The foremost component in the analysis of line transect data is the estimation of a detection function. This function, which is denoted here by $g(x)$, gives the probability of detection of an object at a perpendicular distance x from the transect line. From assumption 1 provided in the preceding section,

$$g(0) = 1.0$$

because the probability is 1.0 that an object with $x = 0$ will be detected. The detection function therefore has a form like that shown in Figure 4.2 for the particular example of a half-normal detection function (the positive half of a normal distribution density function) scaled to have a height of $g(x) = 1.0$ when $x = 0$.

From assumptions 1 to 6, it can be shown that the average probability of detection for an object in the strip of width $2w$ is estimated by

$$\hat{P}_w = 1 / \{w \, \hat{f}(0)\} \tag{4.1}$$

where $f(x)$ denotes the probability density function for the observed distances x. To make use of this equation, the function $f(x)$ is estimated by a curve fitted to the relative frequency histogram of observed x values, and $\hat{f}(0)$ is estimated by the intersection of $f(x)$ with the vertical axis at $x = 0$, as shown

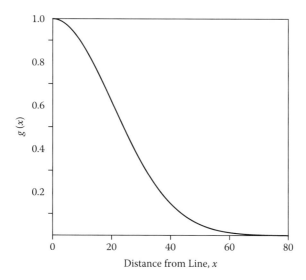

FIGURE 4.2
Graph of the half-normal probability-of-detection curve. Units on the horizontal axis are arbitrary. The height of the curve $g(x)$ has been scaled so that $g(0) = 1.0$.

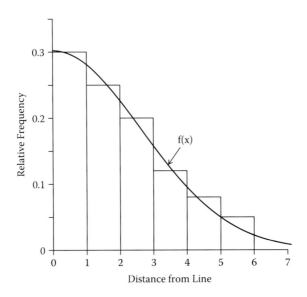

FIGURE 4.3
A curve *f(x)* fitted to the relative frequency histogram for observed distances from objects to a transect line.

in Figure 4.3. Using the previous expression for the estimated probability of detection $\hat{P}_w$, it can be shown that the estimated total number of objects $\hat{N}$ is

$$\hat{N} = n / \hat{P}_w, \tag{4.2}$$

where n is the actual number of animals detected in the study, which is sometimes called the encounter rate.

If n objects are detected in the strip of width $2w$ and length L, then the observed density is $n/(2Lw)$. This is then corrected for visibility bias by dividing by the average probability of detection of objects, as given by Equation (4.1) to obtain

$$\hat{D} = \frac{n/(2Lw)}{1/(w\,\hat{f}(0))} = \frac{n\,\hat{f}(0)}{2L}. \tag{4.3}$$

Note that the width of the transect strip cancels out of this equation. It is therefore acceptable to conduct line transect surveys with no upper limit to the sighting distance. However, it has been found desirable to set an upper limit on w so that a few percent of the most extreme observations are truncated as outliers, using a value of w that can be set after the data are collected.

A more general equation for density estimation can be obtained if objects occur in clusters (see Buckland et al., 2001). The actual number of objects detected in the study n is then interpreted as the estimated expected number of clusters, and the estimated expected cluster size for the population $\hat{E}(s)$ is a multiplier in Equation (4.3), producing

$$\hat{D} = \frac{n \cdot \hat{f}(0) \cdot \hat{E}(s)}{2L}.\tag{4.4}$$

When objects do not occur in clusters (i.e., all objects are detected singly), then $\hat{E}(s) = E(s) = 1$, and Equation (4.4) becomes Equation (4.3).

The choice of the detection function might be important for the estimates obtained. Some nonparametric methods are

a. fitting a curve subjectively by eye,
b. using a Fourier series approximation,
c. using an exponential power series approximation,
d. using an exponential polynomial approximation, and
e. using what is called a key function, with a series adjustment like $g(y) = \text{key}(y)\{1 + \text{series}(y)\}$.

These are described as nonparametric because they do not assume a standard statistical distribution for the distance x of objects from the transect line. Two parametric methods assume that x has

f. a negative exponential distribution and
g. a half-normal distribution.

As noted, sometimes the observed data contain extreme observations at the right of the scale of distances. If this situation occurs, a truncation process is suggested, that is, the removal of observations found a long way from the track line. Buckland et al. (2001) recommend two rules of thumb: The simplest one is to truncate at least 5% of the most extreme distances, and the other alternative involves the estimation of a preliminary detection function $\hat{g}(x)$ with all the distances included and then removing the large distances x such that $\hat{g}(x) \geq 0.15$ for the remaining distances.

It has been shown mathematically and by computer simulation that the detection function can be made up of a mixture of more simple functions that depend on factors such as weather, observer training, and so on, as long as all such functions satisfy the condition that probability of detection is 1 when $x = 0$. For a full discussion on the choice of the detection function and truncation, see the work of Buckland et al. (2001).

EXAMPLE 4.1 Sampling Waterfowl Nests

Anderson and Pospahala (1970) made a study of duck nest density that involved a total of $L = 1600$ miles of transects being walked using a half-width of $w = 8.25$ ft. There were $n = 534$ nests detected within 8 ft, which yielded the observed relative frequency histogram in Figure 4.4. L. McDonald (unpublished manuscript, 2011) has suggested an ingenious "eyeballed" procedure for curve fitting the distribution of observed distances, which is based on Equation (4.1). The key in this method is to let w = number of bars in the histogram. Here, a rough estimate of the height of the function $f(x)$ at the origin is $\hat{f}(0) = 0.1457$ per foot. The probability that a duck nest is detected in this study is estimated as

$$\hat{P}_w = 1/\{w\,\hat{f}(0)\} = 1/\{8 \times 0.1457\} = 0.86 \quad or \quad 86\%.$$

From Equation (4.2), the estimated number of nests is

$$\hat{N} = n/\hat{P}_w = 534/0.86 = 621.$$

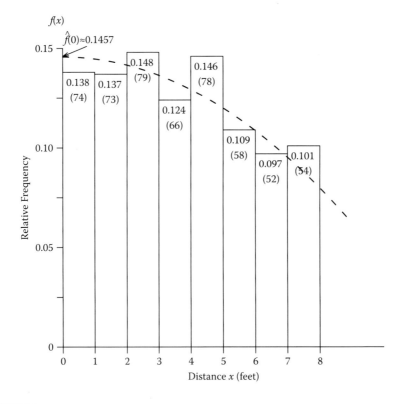

FIGURE 4.4
Relative frequency histogram of right-angle distances to duck nests in the Anderson and Pospahala (1970) study. The proportion of nests in each distance class is reported on the histogram; raw counts of number of nests detected within each interval are in parentheses.

The estimate of duck nest density, Equation (4.3), on this basis is

$$\hat{D} = \frac{(534)(0.1457)}{2(1600)(5280)} = 0.000004605 \text{ nest per square foot,}$$

or 128.4 nests per square mile. Note that $L = 1600 \times 5280$ ft.

Anderson and Pospahala fitted a second-degree polynomial equation to the midpoints of the top of the bars in the histogram. Their analysis procedure was formalized in a slightly different manner, but it is equivalent to the estimation of $f(0)$ by $\hat{f}(0) = 0.1442$, leading to $\hat{D} = 127.1$ nests per square mile.

The most widespread approach used to estimate the probability density function for the observed distances at $x = 0$, $\hat{f}(0)$, dictates the use of any of the parametric and nonparametric methods mentioned. For example, a Fourier series was traditionally applied in older studies of line transect sampling to approximate $f(x)$. Currently, there are plenty of options for density estimation implemented in Distance (Thomas et al., 2009), a program that has become standard for line transect analysis (and point transect analysis; see Section 4.8). Although Distance does not have an explicit option called Fourier series, it can be built with the combination of the uniform key function $1/w$ and an adjustment given by the cosine series expansion,

$$\sum_{j=1}^{m} a_j \cos(j\pi y / w).$$

EXAMPLE 4.2 Sampling Waterfowl Nests (continued)

Assuming a Fourier series to approximating the density function, that the number of detected objects has a Poisson distribution, that sightings of nests are independent events, and that distance data are transformed into eight intervals for analysis, Distance yields $\hat{f}(0) = 0.1477$; hence, $\hat{D} = 130.1$ nests per square mile. The approximate 95% confidence interval for the population density of nests is from 114.0 to 148.5 nests per square mile.

Using Equations (4.1) and (4.3), a more intuitive justification for the estimation formula can be given. The observed density of duck nests is

$$\hat{D} = n/(2Lw) = 534/\{2(1600)(8/5280)\} = 110.14 \text{ nests/square mile.}$$

The average probability of detection of a nest within 8 ft of the transect line is estimated to be about 0.8463 by the formula

$$\hat{P}_w = 1/\{w \, \hat{f}(0)\} = 1/\{8(0.1477)\} = 0.8463.$$

Finally, the observed density adjusted for the visibility bias is

$$\hat{D} = \{n/(2Lw)\}/(\hat{P}w) = 110.14/0.8463 = 130.1 \text{ nests/square mile.}$$

It is necessary to be careful with the units when plotting histograms and doing calculations. The units for the length of transect L must be the width of one class in the histogram. In the duck nest example,

$$\hat{f}(0) = 0.148 / ft, \text{ and } L = (1600)(5280) \text{ ft}.$$

If the width of a class were 0.5 ft, then

$$\hat{f}(0) = 0.148 / 2 = 0.074 \; ft \text{ per } 0.5 \text{ ft},$$

and

$$\hat{D} = (534)(0.074)(2) / \{2(8448000)\} = 0.000004678 \text{ nests/ft}^2.$$

A common mistake made when plotting histograms and fitting the function $f(x)$ occurs when classes for grouped data are of different widths. In this case, the height (and area) of the histogram bars must be adjusted to yield a histogram with total area of 1.0 for all bars. If this is not accomplished, then incorrect impressions will be drawn concerning the detection function.

4.4 Estimation from Sighting Distances and Angles

As noted, it is less satisfactory generally to record sighting distances r and angles θ than it is to record distances from the transect line x, as shown in Figure 4.1. But, if it is for some reason necessary to use r and θ, then estimators of density are available. One approach is Hayne's (1949) method, which assumes a circular flushing envelope.

The basic estimator is

$$\hat{D} = \frac{1}{2L} \sum_{i=1}^{n} \frac{1}{r_i}.$$

Thus suppose that five sighting distances r_i in meters (5, 3, 1, 2, and 4) are made on a transect of length $L = 1000$ m. Then, the reciprocals $1/r_i$ are 0.200, 0.333, 1.000, 0.500, and 0.250, respectively, with sum 2.283, and the estimated density is

$$\hat{D} = \{1 / (2 \times 1000)\}(2.283) = 0.00114 \text{ objects/m}^2.$$

4.5 Estimation of Standard Errors in Line Transect Sampling

Currently, with the versatile options implemented in the Distance program, there are many choices for obtaining the most suitable detection function for the observed distances (e.g., on the basis of goodness of fit or information criteria) and the corresponding estimate of the density and abundance with the data at hand. In the classic distance-sampling setting, the user may choose among four key functions and three series adjustments (see Table 4.1). In addition, there are different options for estimating the standard error of a density estimate using line distances. When there is only one stratum level and one sample, this standard error is estimated in Distance. The standard errors of the number of objects in the sampled area (the encounter rate) and the detection probability are combined to give the standard error of the density estimate. A more general equation for this case, given in Equation (3.68) of Buckland et al. (2001), involves the expected cluster size as a third component of the standard error equation. Among these three components, the standard error of the encounter rate is the most difficult to estimate, and its estimation can be biased when there are few samples. These equations assume a Poisson distribution for the number of objects in the sampled area, and that points are randomly located in the study region. However, it is better if possible not to rely on specific assumptions like this and instead estimate the standard error directly from the results obtained from replicating the sampling process.

True replications of the sampling process should be physically distinct and be located in the study area according to a random procedure that provides an equal chance of detection to all individuals. Given independent replication of a line or set of lines, the density should be estimated for each replication and the standard error of density estimated by the usual standard error of the mean density (weighted by line length if lines vary appreciably in length). For an

TABLE 4.1

Key Functions and Series Adjustments Implemented in Distance when Right Truncation Is Applicable to the Distance Data

Key Function	Form	Series Adjustment	Form
Uniform	$1/w$	Cosine	$\sum_{j=2}^{m} a_j \cos(j\pi y / w)^*$
Half-normal	$e^{-\frac{1}{2}y^2/\sigma^2}$	Simple polynomial	$\sum_{j=2}^{m} a_j (y/w)^{2j}$
Hazard rate	$1 - e^{-(y/\sigma)^{-b}}$	Hermite polynomial	$\sum_{j=2}^{m} a_j H_{2j}(y/w)$
Negative exponential	e^{-ay}		

Source: Adapted from the tables given in Chapter 8 of the *Distance User's Guide* by Thomas, L., Laake, J.L., Rexstad, E., Strindberg, S., Marques, F.F.C., Buckland, S.T., Borchers, D.L., Anderson, D.R., Burnham, K.P., Burt, M.L., Hedley, S.L., Pollard, J.H., Bishop, J.R.B., and Marques, T.A. (2009). *Distance 6.0.* Release 2. Research Unit for Wildlife Population Assessment, University of St. Andrews, UK (http://www.ruwpa.st-and.ac.uk/distance/).

Note: Here, y is distance, w is the truncation distance, and σ, a, and b are model parameters. Hermite polynomial functions $H_x(y/w)$ are defined in the book by Stuart and Ord (1987, pp. 220–227).

* When a uniform key function is used, the summation is from $j = 1$ to m.

account of the methods available for the estimation of the encounter rate standard error using replicated transect lines, see the article by Fewster et al. (2009) and Chapter 8 in the Distance user's guide. If there are not enough detections on independent replications of lines or systematic sets of lines to obtain independent estimates of the density, then it is recommended to use bootstrapping to estimate the variance by resampling transect lines.

4.6 Size-Biased Line Transect Surveys

Drummer and McDonald (1987) considered the use of line transect surveys for objects when the size of an object influences the probability of detection. For example, Drummer et al. (1990) considered applications to populations of clusters of individuals rather than single individuals (for the sea otter *Enhydra lutris*) and to pellet groups of barren ground caribou (*Rangifer tarandus granti*). If group size influences the probability of detection of a group, then the observed mean group size is biased and will lead to overestimation of the density of individuals. See the work of Buckland et al. (2001, 2004) for more possibilities about data analysis under these conditions.

4.7 Probability of Detection on the Line of Less than One

Because of observer bias and other problems, the assumption of 100% detection of items on the transect line might not be reasonable. Quang and Lanctot (1991) extended line transect sampling theory to the case when there is less than 100% detection on the line and detection function is unimodal. The assumption that there is 100% probability of detection on the line is replaced by the assumption that a line of perfect detection exists parallel to the transect line at some unknown distance away. Application of the technique is made to the estimation of common and pacific loon (*Gavia pacifica* and *Gavia immer*) in the Yukon Flats National Wildlife Refuge from aerial surveys. It is difficult to guarantee perfect detection on the inside edge of the strip because of the speed at which the aircraft passes a loon. The assumption of perfect detection at a certain distance from the inside edge is more reasonable. If the assumption of 100% detection on the line, on the inside edge of the strip, or at an unknown distance away is violated, then density estimates obtained by the line transect theory are biased underestimates. In this case, the procedure is conservative because the population size is underestimated by some amount. One possibility is then to use double observers for some or all of the transect sampling. In that case, items can be detected by observer A only, observer B only, or both observers. It is then possible to estimate the probability of detection by one or both observers at different distances from the line. One approach was provided by Manly et al. (1996), who used it on data from aerial surveys of polar bears in the Arctic. This approach is easily modified to allow the probabilities of detection to depend on covariates. Other approaches with double observers were discussed by Buckland et al. (2001, 2004).

EXAMPLE 4.3 Sperm Whale Sampling

In the Gulf of California, Mexico, the density and abundance of sperm whales *Physeter macrocephalus* were estimated in spring 2005 by setting up line transect surveys (Díaz-Gamboa, 2009). The observation platform was 6.6 m high from a vessel at 8.3 knots speed with two main observers covering 180° visual range to the front using 7 × 50 binoculars equipped with reticle and compass and an independent observer covering 360° of visual range to report sightings undetected by the main observers. Observations were made at Beaufort 3 sea state or less to reduce bias. The reticle and sighting angle degree were reported on each sighting to determine the perpendicular distance from an algorithm that converted radial to linear distance at the proper latitude. The geographic position and transect course were registered every minute. The study area was divided into northern, central, and southern strata (Figure 4.5).

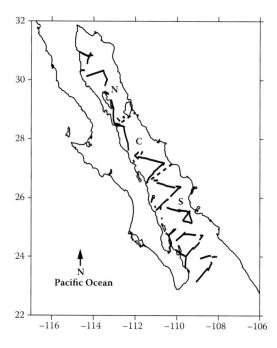

FIGURE 4.5
Sighting effort for sperm whale sampling from the Gulf of California, Mexico, using line transect sampling. Transects were grouped in three strata: northern (N), central (C), and southern (S) Gulf of California.

The effort in the northern stratum was 493.56 km, in the central stratum it was 406.01 km, and in the southern stratum it was 1199.81 km, with a total effort of 2099.38 km. The detection function and cluster size were estimated for all strata, but the encounter rate was estimated separately for each stratum. A cluster was considered when two or more individuals were sighted at a distance of less than a body length from each other. The density, abundance, variance, and confidence intervals were estimated in Distance (Thomas et al., 2009). The hazard rate, half-normal, and uniform models with cosine, simple polynomial, and Hermite polynomial series expansions were used, and the one with the lowest value of Akaike's information criterion (AIC) was selected. It was assumed that not all the sperm whales were detected at zero distance $g(0)$, and the sighting information by independent observers was used to estimate the proportion of individuals not detected by the main observers.

The estimated value of $g(0)$ was 0.81 with a standard error of 0.18. This estimate improved previous sperm whale density estimates. Previously, it was customary to underestimate it because a fraction of animals is not usually available to the main observers because of the typical sperm whale diving behavior (Barlow and Sexton, 1996). A total of 132 sperm whales were observed in 89 sightings distributed only in the central and southern strata. The mean cluster size was 1.4 sperm whales, and there was a standard error of 0.07. Given the frequency of perpendicular distances from sperm whale sightings, the distribution of these distances was truncated at 5.5 km to discard all the observations beyond that distance, as shown in Figure 4.6. The uniform model with cosine series expansion was selected. The estimated value of $f(0)$ was 0.344 km^{-1} with percentage coefficient of variation (% CV) = 4.46; the sperm whale density and abundance estimates by strata are shown in Table 4.2.

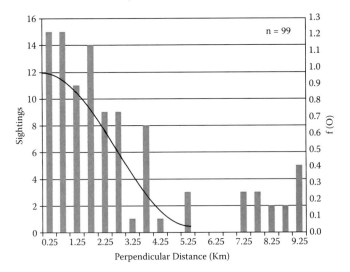

FIGURE 4.6
Histogram of perpendicular distances of sperm whale sightings in the Gulf of California, sampled by line transect surveys. The fitted detection function is also shown (see text).

TABLE 4.2

Sperm Whale Density and Abundance Estimates in Spring 2005 in the Gulf of California, Mexico

Strata	Northern				Central				Southern			
Parameter	E	% CV	LCL	UCL	E	% CV	LCL	UCL	E	% CV	LCL	UCL
n/L	0	—	—	—	0.071	95.25	0.013	0.389	0.038	41.90	0.017	0.086
D	0	—	—	—	0.021	102.11	0.003	0.121	0.011	55.77	0.004	0.031
N	0	—	—	—	595	102.11	103	3,427	987	55.77	353	2,759

Note: E, estimate; LCL, 95% Lower Confidence Limit; UCL, 95% Upper Confidence Limit; n/L, encounter rate, No. $\cdot$ km^{-1}; D, density, No. $\cdot$ km^{-2}; N, abundance.

4.8 Point Transect Sampling

Point transects are like line transects for which the lines have zero length. What happens is that a number of points are chosen within the study area, and each point is visited. The objects detected from each point are then recorded, with their distances from the points, as shown in Figure 4.7. This sampling method is called variable circular plot sampling by ornithologists, who seem to be the main users of such methods. Information about the analysis of data from such studies was provided by Buckland et al. (2001).

4.9 Software for Line and Point Transect Sampling and Estimation

Distance, the standard program for line and point transect sampling and estimation, is available free on the Internet from the University of St. Andrews in Scotland (Thomas et al., 2009), and an updated description of the history and procedures currently implemented in this program has been published by the original authors and contributors (Thomas et al., 2010). The most recent version of Distance includes three analysis engines: the conventional or classic distance sampling (line transect and point transect), multiple covariate distance sampling, and mark-recapture distance sampling (MRDS; Borchers et al., 1988). This last method is useful when sampling involves double observers (Manly et al., 1996; Buckland et al., 2010), but it can only run in Distance in conjunction with routines written in the R language. To

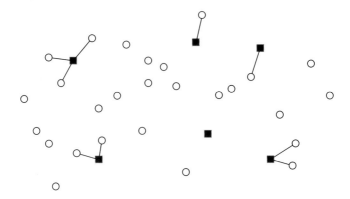

FIGURE 4.7
An illustration of point transect sampling, with six sample points (■) and from none to three objects (O) detected.

use the full facilities of the program Distance, a 362-page user's manual that comes with the program should be read.

Calculations of different parameter estimates in line and point transect sampling can also be performed in three recently published R packages: unmarked (Fiske and Chandler, 2011), Rdistance (McDonald, 2012), and mrds (Laake et al., 2013). Unmarked is an R package that fits a variety of hierarchical models to data collected using survey methods such as point counts, site occupancy sampling, line transect sampling, removal sampling, and double-observer sampling. For the most part, this package performs a subset of the analyses available in the Distance program. Rdistance is able to run all the procedures indicated in Distance as classic distance sampling, but it also includes routines to fit a gamma distance function, and the uniform distance function of Rdistance is parameterized differently. The mrds package analyzes single- or double-observer distance-sampling data for line or point sampling. A simpler interface of mrds, called Distance (homonym of the stand-alone program; Miller 2013), has also been created in R to facilitate the estimation process of the detection function and the abundance/density of objects in line and point transect sampling.

5

Removal and Change-in-Ratio Methods

Lyman McDonald and Bryan Manly

5.1 Introduction

This chapter covers two methods for estimating the size of animal populations. The first is the removal method, which relies on the capture and removal of animals where the number of animals removed has a noticeable impact on the numbers found in later samples. Alternatively, the methods apply if animals can be uniquely marked or tagged and the proportion of unmarked animals found in later samples decreases substantially.

The second is the change-in-ratio method, which relies on the removal of some of one type of animal (e.g., males), leading to a substantial change in the proportion of that type in the population and in later samples. Alternatively, the animals can be uniquely marked and tagged so that the proportion of unmarked animals of one type (e.g., males or juveniles) in later samples changes.

Clearly, the removal method is related to the methods covered in further chapters on mark-recapture sampling, and data can be analyzed by the more complex methods presented in those chapters. In this chapter, we recommend analysis using the computer software program CAPTURE at the interactive website http://www.mbr-pwrc.usgs.gov/software/capture.html (White et al., 1978, 1982; Otis et al., 1978; Rexstad and Burnham, 1991). However, we also present a linear regression method to give an intuitive understanding of the information in the data. The regression method is simpler to understand and will provide approximately the same results as the more complex method.

5.2 Removal Method

The simplest removal method involves taking a series of samples from an animal population and removing or marking the animals that are captured.

With this sampling scheme, the remaining number of unmarked animals in the population will increasingly reduce over time because of the removals or marking until (if sampling is continued long enough) there are no animals or unmarked animals left. At that stage, the total number of animals removed or marked will equal the initial population size, providing that the number of animals has not changed for reasons other than the removals.

Removal sampling carried on long enough can therefore provide an exact value for the population size. However, in practice the main value of the method is that the results after a few samples have been taken can be used to extrapolate to the results that would be obtained from taking many samples and hence produce an estimate of the population size. In the rest of this chapter, we use the phrase *removal of animals* with the understanding that animals can be effectively removed by marking all animals caught (i.e., animals do not have to be physically removed from the population).

We present two removal models. The simplest model is equivalent to the behavior model, model M_b or the Zippin model (Zippin, 1956, 1958) for a closed population, because the probability of capture in later samples changes to zero when an animal is removed. The probability of capture is not changing because of an animal's behavior, such as trap avoidance, but rather because the animal is removed. The estimation of population size using the Zippin model relies on the following assumptions, which ensure that when a sample is taken the expected number of captures is proportional to the number of individuals that have not yet been removed:

a. When a sample is taken, each available individual in a population has the same chance of being captured (there is no heterogeneity in capture probability).

b. The population is closed so that the size remains constant over the sampling period, apart from the removal losses.

c. Catching one individual does not change the probability of catching other individuals.

The assumptions are generalized in the following material in a different model to allow each animal to be removed with a different probability.

As an example, suppose that the fish population in a lake is sampled with equal effort for 5 days with 65, 43, 34, 18, and 12 fish removed on the successive days. It is a reasonable assumption that the reduction in numbers is caused by the removals, and that most fish were taken or marked by day 5. On this basis, the total number of fish that were originally in the lake can be estimated by extrapolating the number of fish sampled beyond day 5 to obtain a total that is about 180, as shown in Figure 5.1. Using the computer program CAPTURE to carry out the calculations for the Zippin model, the population estimate under the assumptions of the Zippin model is 196 with

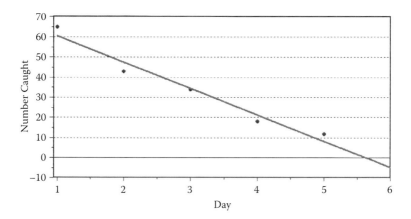

FIGURE 5.1
How the removal method reduces the population as more fish are removed from the population on each day. The observed numbers removed are shown for each day (•), and the line is a linear regression fitted to these numbers. The regression line suggests that on day 5 there were about eight fish left in the population. The estimated total population size was therefore about $65 + 43 + 34 + 18 + 12 + 8 = 180$.

standard error 9.9, suggesting that on day 5 there were 24 animals left in the study area.

For a second example, assume that the assumptions hold, the population size is $N = 100$, and 20% of the remaining population is expected to be caught on each occasion. The expected size of the catch on the first occasions is $z_1 = 20$. On the second occasion, 20% of the remaining 80 or $z_2 = 16$ animals are expected to be captured. On the third occasion, the model predicts that $z_3 = 12.8$ animals would be captured. This is an impossible value, so sample data and modeled values cannot agree exactly. Still, continuing, the expected catch sizes on the fourth and fifth occasions are $z_4 = 10.24$ and $z_5 = 8.192$, respectively. At the end of five capture occasions, the expected value of the number of animals remaining in the population is $100(1 - 0.2)^5 = 100(0.8)^5 = 32.768$, that is, about 33 animals. The total catch is then expected to be about $(z_1 + z_2 + \ldots + z_5) = 67$ animals. Generalizing these results, we can develop a model relating the population size to the probability of capture and the total catch.

Assuming that assumptions a, b, and c hold, let z_i denote the catch in the ith sample, N denote the true population size, p denote the expected proportion of the available population captured in a sample, and n denote the number of samples taken. Then, the expected numbers in the successive catches are $z_1 = Np$, $z_2 = N(1 - p)p$, $z_3 = N(1 - p)^2p$, and in general $z_n = N(1 - p)^{n-1}p$.

Another approximate relationship is obtained by noting that the expected number remaining in the population after the nth removal is $N(1 - p)^n$. This plus the number removed adds to N, so that

$$N \approx (\text{Number removed}) + N(1 - p)^n,$$

or

$$N \approx (z_1 + z_2 + \ldots + z_n) + N(1 - p)^n.$$

Solving this equation for N, we have a model for the relationship between N, total catch, and p, which is

$$N \approx (z_1 + z_2 + \ldots + z_n)/\{1 - (1 - p)^n\}. \qquad (5.1)$$

Continuing the numerical example and assuming the total catch is 67, Equation (5.1) yields the result that

$$N \approx (67)/\{1 - (0.8)^5)\} = 99.7.$$

Hence, with real-life data Equation (5.1) gives an estimate of N.

The expected catch in the ith sample is $z_i = N(1 - p)^{i-1}p$. Taking the logarithm of both sides of this equation yields the approximate linear relationship

$$\log_e(z_i) \approx \log_e(N) + \log_e(p) + (i - 1) \log_e(1 - p),$$

or

$$\log_e(z_i) \approx a + (i - 1)b,$$

so that

$$y_i \approx a + bx_i, \qquad (5.2)$$

where $y_i = \log_e(z_i)$, $x_i = i - 1$, and the slope of the line is $b = \log_e(1 - p)$. This equation therefore gives a model for the natural logarithm of the number removed in the ith sample.

On the basis of Equations (5.1) and (5.2), Soms (1985) proposed the following method for estimating N and p. First, the values $y_i = \log_e(z_i)$ and $x_i = i - 1$ are calculated for each of the sample, $i = 1, 2, \ldots, n$. Next, the equation $y_i = a + bx_i$ is fitted by ordinary linear regression methods to obtain an estimate of the slope b. Then, because b in the regression equation is $b = \log_e(1 - p)$ in Equation (5.2), an estimate of p can be found by solving the equation $\log_e(1 - \hat{p}) = b$ to give

$$\hat{p} = 1 - \exp(b). \qquad (5.3)$$

Finally, Equation (5.1) suggests the estimator

$$\hat{N} = (z_1 + z_2 + \ldots + z_n)/\left\{1 - (1 - \hat{p})^n\right\},$$

or

$$\hat{N} = (\text{Total catch})/\left\{1 - (1 - \hat{p})^n\right\} \tag{5.4}$$

for N.

Soms (1985) has shown that Equations (5.3) and (5.4) yield estimates that are about as good as those obtained from the more complicated method developed by Zippin (1956, 1958). Soms also provided the following equations for the variances of the regression estimators, noting that they appear to give satisfactory results providing that the population size N is about 200 or more:

$$\text{Var}(\hat{N}) \approx N\left\{q^n / (1 - q^n)\right\}\left\{1 + n^2 q^n / (1 - q^n) \sum_{i=1}^{n} c_i^2 / s_i\right\}, \tag{5.5}$$

and

$$\text{Var}(\hat{p}) \approx q^2 \sum_{i=1}^{n} c_i^2 / (N s_i), \tag{5.6}$$

where $q = 1 - p$, $c_i = \{i - (n + 1)/2\}/\{n(n^2 - 1)/12\}$, and $s_i = p(1 - p)^{i-1}$.

Zippin (1956) included an example in which small mammals were marked in a 3-night trapping program; the captures were $z_1 = 165$, $z_2 = 101$, and $z_3 = 54$. Based on approximations for the maximum likelihood estimates available in the 1950s, he estimated $\hat{N} = 400$, with standard error 26.3, a 95% confidence interval of 347 to 453, and an estimated probability of capture of $\hat{p} = 0.4253$.

With Soms's method, a linear regression line is fitted to the data ($y_i = \log_e(z_i)$, $x_i = i - 1$): (5.106, 0), (4.615, 1), and (3.989, 2). The equation of the line is $y = 5.1285 - 0.5585x$, so that the estimated proportion of animals removed each night is $\hat{p} = 1 - \exp(-0.5585) = 0.4279$, which is close to Zippin's (1956) estimate. A total of 320 animals were captured, and the total population size was estimated to have been 320, divided by the probability that an animal will be captured during the three trapping nights, $\hat{N} = 320/\{1 - (1 - \hat{p})^3\} = 320/(0.8128) = 394$. Using Equation (5.5), the estimated variance of $\hat{N}$ is 812.37, so that the estimated standard error is $\sqrt{812.37} = 28.5$, and the approximate 95% confidence interval, $394 \pm (1.96)(28.5)$, has a lower limit of 338 and upper limit of 450.

Using the interactive program CAPTURE, the maximum likelihood estimates of the proportion of animals removed each night are $\hat{p} = 0.4253$, and $\hat{N} = 394$ with standard error $= 22.6$, with a 95% confidence interval for N of 362 to 453. The population estimates from Zippin's (1956) article (400) and the program CAPTURE (394) agree favorably with those provided by Soms's linear regression method (394); however, the confidence limits provided by

TABLE 5.1

Comparison of Population Estimates Provided by Two Approximate
Methods Compared to Estimates Provided by CAPTURE Program

	$\hat{N}$	95% Confidence Interval	
		Lower Limit	**Upper Limit**
Zippin (1956)	400	347	453
Soms (1985)	394	338	450
CAPTURE	394	362	453

the CAPTURE program were apparently adjusted for nonnormality of the
estimate (Table 5.1).

EXAMPLE 5.1 Estimation of a Crab Spider Population Size

Soms (1985) illustrated his method of estimation using data on crab spiders. In one case, six samples were taken, and 46, 29, 36, 22, 26, and 23
spiders were removed. The calculations for this example are shown in
Table 5.2. The proportion removed on each sampling occasion is estimated as 0.115, with standard error 0.040, and the estimate of the population size is 350, with standard error 87.6. Soms also described a test for
the goodness of fit of the removal sampling model. With the spider data,
it provides a chi-squared value of 4.06, with four degrees of freedom.
This is not significantly large and therefore gives no reason to doubt the
assumptions made.

Running the interactive CAPTURE program shows that the maximum
likelihood estimate of the population size is $\hat{N} = 324$ with standard
error = 69.03 and with the approximate 95% confidence interval 240 to

TABLE 5.2

Results for the Removal Sampling of Crab Spiders

		Regression Data		Quantities Needed for Variance Calculations		
Sample i	Removed z	X	y	c	s	c^2/s
1	46	0	3.829	−0.143	0.115	0.177
2	29	1	3.367	−0.086	0.102	0.072
3	36	2	3.584	−0.029	0.090	0.009
4	22	3	3.091	0.029	0.080	0.010
5	26	4	3.258	0.086	0.071	0.104
6	23	5	3.135	0.143	0.062	0.327
Total	182					0.699

Note: The fitted regression equation is $y = 3.683 - 0.122x$. Hence, $\hat{p} = 1 - \exp(-0.122) = 0.115$,
$\hat{q} = 0.885$, and $\hat{N} = 182/(1 - 0.885^6) = 350.3$. From Equation (5.5), with N and q replaced by
their estimates $\text{Var}(\hat{N}) \approx 7801.2$ and $\text{SE}(\hat{N}) = 87.6$. Similarly, from Equation (5.6), $\text{Var}(\hat{p}) \approx$
0.0016 and $\text{SE}(\hat{p}) \approx 0.040$.

532. By comparison, the regression method gives $\hat{N} = 350$ with an estimated standard error of 87.6 (Table 5.2).

The CAPTURE program also provides the possibility of relaxing the assumption that capture probabilities are the same at different sample times. The heterogeneous removal model is denoted by M_{bh} and is widely used, especially in fisheries science. Using this model, the estimated population size is the same as that for the Zippin model, with the same standard error.

5.3 The Change-in-Ratio Method

The principle behind the change-in-ratio method is that if a population contains two recognizable types of individual, such as males and females, with estimated proportions at some initial point in time, and the proportions are estimated again after a fixed number of one of the two types of individual have been removed, then it is possible to estimate the population size at the initial point in time on the assumption that population changes are caused only by the removals. Selective harvest of fish and wildlife for certain sexes or desired size often results in changes in ratios of the remaining individuals so that the methods can be applied. For example, a crab fishery may allow only males of a certain size to be harvested. Ratios of males to females or large males to small males are expected to change after harvest of a known number of large males. It is not necessary for animals to be removed from the population; rather, individuals can be marked or tagged. For example, the ratio of unmarked males to females would be expected to change as a known number of males are captured and marked. The first application seems to have been described by Kelker (1940, 1944) in terms of changes in the sex ratio resulting from a differential kill of male and female deer in a harvested population.

The simplest case involves a population partitioned into two subclasses. Surveys are conducted to estimate the subclass ratios at the beginning of the study. Known numbers of the subclasses are then removed or marked. Then, a second survey is conducted to reestimate the subclass ratios. In the following, we refer to removal of animals; however, it is to be understood that animals are effectively removed if captured and marked.

The critical assumption required for estimation of the population sizes is that the probability of encountering the two subclasses is the same for the first and second samples. However, the probability of encounter can change from the first to the second. Two other assumptions are also necessary: The number of animals removed (or marked) should be known, although in practice precise estimates can be used. Finally, the population should be closed, that is, there is no recruitment, immigration, emigration, or unknown mortality. Consequently, the time between the two sample surveys should be short and during a season when animals are not migrating.

To understand the mathematical basis of the method, suppose that the two types of individuals are labeled X and Y. For two sample survey times, $i = 1$ and 2, let

x_i = the number of type X individuals in the population at time i,

y_i = the number of type Y individuals in the population at time i,

$N_i = x_i + y_i$, the total population size at time i,

$p_i = x_i/N_i$, the fraction of the population that is type X at time i,

$r_x = x_1 - x_2$, the number of type X individuals removed between times 1 and 2,

$r_y = y_1 - y_2$, the number of type Y individuals removed between times 1 and 2, and

$r = r_x + r_y = N_1 - N_2$ is the total number of individuals removed (which follows from the previous definitions).

This means that the proportion of type X after the removal is $p_2 = (x_1 - r_x)/(N_1 - r) = (p_1 N_1 - r_x)/(N_1 - r)$. Solving for N_1 gives $N_1 = (r_x - r p_2)/(p_1 - p_2)$. Hence, we can obtain estimates of p_1 and p_2 from a sample survey of the population before and after removals of r_x and r_y animals. If $\hat{p}_1$ and $\hat{p}_2$ are estimates of p_1 and p_2, respectively, then N_1, x_1, and N_2 can be estimated by

$$\hat{N}_1 = (r_x - r\hat{p}_2)/(\hat{p}_1 - \hat{p}_2), \tag{5.7}$$

$$\hat{x}_1 = \hat{p}_1 \hat{N}_1, \tag{5.8}$$

and

$$\hat{N}_2 = \hat{N}_1 - r, \tag{5.9}$$

respectively, assuming that r_x and r_y are known. From these, other parameters can be estimated. For example, the number of type Y animals at time 1 can be estimated by subtracting Equation (5.8) from Equation (5.7).

This formulation of the situation allows for one or both of the two types of individuals to be removed. It also permits negative removals, so that new individuals of one or both types can be added. Finally, the exploitation rate, $u = r/N_1$, of harvested animals can be estimated, which is important for management of a population.

Assuming that $\hat{p}_1$ and $\hat{p}_2$ are estimated by the proportions seen in samples of size n_1 and n_2, taken at times 1 and 2, respectively, the variances of $\hat{N}_1$ and $\hat{x}_1$, respectively, are given approximately by the equations

$$\mathrm{Var}(\hat{N}_1) \approx \{N_1^2\,\mathrm{Var}(\hat{p}_1) + N_2^2\,\mathrm{Var}(\hat{p}_2)\}/(p_1 - p_2)^2, \tag{5.10}$$

and

$$\mathrm{Var}(\hat{x}_1) \approx \{(N_1 p_2)^2\,\mathrm{Var}(\hat{p}_1) + (N_2 p_1)^2\,\mathrm{Var}(\hat{p}_2)\}/(p_1 - p_2)^2, \tag{5.11}$$

where

$$\mathrm{Var}(\hat{p}_i) = \{p_i(1 - p_i)/n_i\}(1 - n_i/N_i). \tag{5.12}$$

These variances can be estimated by substituting estimates for true parameter values, as necessary. The variance of $\hat{N}_1$ is the same as that of $\hat{N}_2$, and the variance of $\hat{x}_1$ is the same as that of $\hat{x}_2$. In general, the change-in-ratio method gives estimates with good precision if the number of animals removed is large and highly selective for one of the subclasses. Equivalently, in a capture-and-marking study, a large number of animals in one of the subclasses must be marked. Spurious results can be obtained if the ratios of the two subclasses are little changed between the two sample surveys.

Approximate confidence limits can be calculated in the usual way as $\hat{N}_1 \pm z_{\alpha/2}\widehat{\mathrm{SE}}(\hat{N}_1)$ and $\hat{x}_1 \pm z_{\alpha/2}\widehat{\mathrm{SE}}(\hat{x}_1)$, where the estimated standard errors are the square roots of the estimated variances, and $z_{\alpha/2}$ is the value exceeded with probability $\alpha/2$ for the standard normal distribution. Better confidence limits for small samples were reviewed by Seber (1982, p. 363).

EXAMPLE 5.2 Estimating a Mule Deer Population Size

Rasmussen and Doman (1943) described how a mule deer population near Logan, Utah, suffered a severe loss during the 1938–1939 winter. Counts before the loss gave an estimated proportion $\hat{p}_1 = 0.4536$ of fawns, and after the loss the estimated proportion was $\hat{p}_2 = 0.3464$. A complete survey of the area disclosed $r_x = 248$ dead fawns and $r_y = 60$ dead adults.

From these data, Equation (5.7) yields

$$\hat{N}_1 = (248 - 308 \times 0.3464)/(0.4536 - 0.3464) = 1318.2 \text{ deer}$$

and Equation (6.8) yields

$$\hat{x}_1 = 0.4536 \times 1318.2 = 597.9 \text{ fawns.}$$

Then, there is an estimated number of

$$\hat{y}_1 = 1318.2 - 597.9 = 720.3 \text{ adults,}$$

and

$$\hat{N}_2 = 1318.2 - 308 = 1010.2 \text{ deer.}$$

Assume that the sample sizes used to estimate $\hat{p}_1$ and $\hat{p}_2$ were $n_1 = n_2 = 400$. Then, Equation (5.12) gives estimated variances of

$$\widehat{\text{Var}}(\hat{p}_1) = \{0.4536(1 - 0.4536)/400\}(1 - 400/1318.2) = 0.0004316$$

and

$$\widehat{\text{Var}}(\hat{p}_2) = \{0.3464(1 - 0.3464)/400\}(1 - 400/1010.2) = 0.0003419.$$

Also, from Equation (5.10),

$$\widehat{\text{Var}}(\hat{N}_1) = (1318.2^2 \times 0.0004316 + 1010.2^2 \times 0.0003419)/(0.4536 - 0.3464)^2$$

$$= 95602.9$$

giving $\widehat{\text{SE}}(\hat{N}_1) = \sqrt{95602.9} = 309.2$. An approximate 95% confidence interval for the population size before the severe loss is then $1318.2 \pm 1.96 \times 309.2$ or 712 to 1924. These limits are rather wide even assuming large samples for estimating the population proportion of fawns.

Using the fact that the variance of $(\hat{N}_1)$ is the same as that of $(\hat{N}_2)$ and the variance of $(\hat{x}_1)$ is the same as that of $(\hat{x}_2)$, approximate confidence intervals can be obtained for the initial number of fawns, the final number of fawns, and the final population size. Treating adults as type X animals also allows variance, standard error, and confidence limits to be obtained for the initial number of adults. However, these calculations are not carried further here.

5.4 Relationship between Change-in-Ratio and Mark-Recapture Methods

Assume we capture, mark, and release n_1 animals at time 1 in a closed population of size N_1. At time 2, assume n_2 animals are captured, of which m are marked. The Lincoln–Petersen estimate of the size of the population presented in Chapter 7 is then $\hat{N}_1 = n_1 n_2 / m$. Viewed as a change-in-ratio study, let X denote marked animals and Y denote unmarked animals. At time 1, $x_1 = 0$, $p_1 = 0$, and $y_1 = N_1$ because there are no marked animals in the population. However, there is an addition of n_1 marked animals, leading to a negative removal of $r_x = -n_1$ and a positive removal of unmarked animals, $r_y = n_1$. Finally, the total removed is $r = r_x + r_y = -n_1 + n_1 = 0$. Substituting into the change-in-ratio formula for the initial population size, we have $\hat{N}_1 = (r_x - r\hat{p}_2)/(\hat{p}_1 - \hat{p}_2) = (-n_1)/(-\hat{p}_2) = n_1/\hat{p}_2$. However, the proportion of marked animals at time 2 is estimated by $\hat{p}_2 = m/n_2$, leading to the Lincoln–Petersen estimate $\hat{N}_1 = n_1 n_2/m$. Variances can be computed from either the equations given in Chapter 7 or the ones presented here.

Pollock et al. (1985) have pointed out that a potential practical problem with the use of the change-in-ratio method is that the two types of individual may not be equally likely to be observed in a sample from the population, with the result that $\hat{p}_1$ and $\hat{p}_2$ will be biased estimators. They noted, however, that if the removals are of only one type of animal, then the number of this type can be estimated without bias irrespective of whether the two types are equally visible and discussed how a three-sample design can be used to estimate the numbers of both types of animal. The estimation equations are quite straightforward if one type of animal is removed between the times of sample 1 and sample 2, and then the other type of animal is removed between the times of sample 2 and sample 3. Removals of some of both types between sample 1 and 2 and then the further removal of some of both types between sample 2 and sample 3 can also in principle be allowed, although estimation of population sizes is then somewhat more complicated.

An interesting application of the three-sample change-in-ratio method was described by Lancia et al. (1988) as part of a harvest strategy for populations such as deer for which harvests are well controlled and hunting seasons are short. They showed that taking two separate single-sex harvests together with three samples of deer provides a means of achieving the harvest quota and estimating the population size.

Application of change-in-ratio methods to multiple time periods using a software program called USER was covered by Skalski and Millspaugh (2006). They discussed multiclass and sequential methods and illustrated how the USER program can be used to calculate maximum likelihood estimates. Other extensions of the change-in-ratio method to multiclass situations and a combination of change in ratio with effort information were covered by Udevitz and Pollock (1991, 1995). Computer programs for analysis of these extended methods are provided using SAS statistical software (http://alaska.usgs.gov/science/biology/biometrics/cir01/using_sas.php).

6

Plotless Sampling

Jorge Navarro

6.1 Introduction

Sampling methods have been developed for studying different properties (e.g., density, spatial pattern) of sets of points or items (e.g., trees, nests, etc.) distributed within an area of space. In many instances, sampling is the only solution as it is not possible to map or census all the points in the area; consequently, the field ecologist is constrained to select points from subregions within the total area. Diggle (2003) calls the sampling methods involved in this last case sparse sampling methods to distinguish them from those concerning mapped populations within an area. In general, sparse sampling methods can be divided into two different classes, depending on the way points within a subregion are selected. One method is quadrat or plot sampling, and the other method is plotless sampling. Although quadrat sampling has always been a popular field technique in forestry and plant ecology, plotless sampling techniques have also been applied by foresters and plant ecologists when rapid estimates of the density of plants in a large region are needed. In fact, plotless sampling is often considerably more efficient than quadrat sampling because searching for all the items in a quadrat might be time consuming.

Traditional plotless sampling methods are based on the measurement of distances from random points to the nearest item and distances from the nearest item to its nearest (or kth-nearest) neighbor (Kleinn and Vilčko, 2006), which is why these procedures are also known by the name of distance sampling. However, the term *plotless* is preferred here because distance sampling may also refer to line or point transect sampling methods (Buckland et al., 2001), like those described in Chapter 4.

Plotless sampling methods are useful if the interest is only in the density of certain distinct items (e.g., trees above a certain size), it is relatively easy in the field to lay out a large grid of systematic points that are separated by sufficient distance so that results on neighboring points can be considered independent or to choose a number of randomly located points, and it is relatively easy in the field to find and measure the distance to the closest item, then find and measure the distance to the nearest (or second-nearest, third-nearest, etc.) neighbor of the item.

Numerous procedures have been devised to estimate the density (number per unit area) of distinct items in a study area using plotless sampling. These include the point-centered quarter method (Cottam, 1947) by which the distances from a random point to the closest individual in each 90° quadrant are measured. However, the density estimates from the point-centered quarter method and most variations of distance methods are biased if the spatial pattern is not random. In this chapter, only two plotless sampling procedures are considered: *T*-square sampling and wandering-quarter sampling.

6.2 *T*-Square Sampling

With the *T*-square method a random (or systematic) sample of points is located in the study area with the distances between points large enough for them to be considered independent. At each point, two distances are measured, as shown in Figure 6.1. The first distance x_i is from the point P to the nearest item I, and the second distance z_i is from the item I to its nearest neighbor.

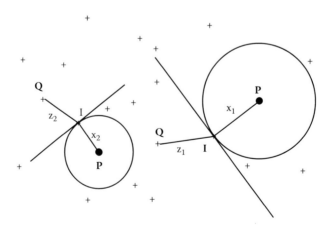

FIGURE 6.1
T-square sampling. Two random points P are located in the study area, and the distances x_i to the closest item I are measured. The distance z_i from the item I to its nearest neighbor Q is then measured subject to the condition that the angle from P to I to Q is more than 90°.

Krebs (1999) has reviewed tests for a random distribution of items and recommends one proposed by Hines and O'Hara Hines (1979). The test statistic is then

$$h_T = 2n[2\Sigma(x_i)^2 + \Sigma(z_i)^2]/[2^{\frac{1}{2}} \Sigma x_i + \Sigma z_i]^2$$

where n denotes the number of points in the sample. The test statistic is compared to the critical values shown in Table 6.1, with the table entered with

TABLE 6.1

Critical Values for the Hines and O'Hara Hines (1979) Test Statistic h_T

α	Regular Alternative				Aggregate Alternative			
m	0.005	0.01	0.025	0.05	0.05	0.025	0.01	0.005
10	1.0340	1.0488	1.0719	1.0932	1.4593	1.5211	1.6054	1.6727
12	1.0501	1.0644	1.0865	1.1069	1.4472	1.5025	1.5769	1.6354
14	1.0632	1.0769	1.0983	1.1178	1.4368	1.4872	1.5540	1.6060
16	1.0740	1.0873	1.1080	1.1268	1.4280	1.4743	1.5352	1.5821
18	1.0832	1.0962	1.1162	1.1344	1.4203	1.4633	1.5195	1.5623
20	1.0912	1.1038	1.1232	1.1409	1.4136	1.4539	1.5061	1.5456
22	1.0982	1.1105	1.1293	1.1465	1.4078	1.4456	1.4945	1.5313
24	1.1044	1.1164	1.1348	1.1515	1.4025	1.4384	1.4844	1.5189
26	1.1099	1.1216	1.1396	1.1559	1.3978	1.4319	1.4755	1.5080
28	1.1149	1.1264	1.1439	1.1598	1.3936	1.4261	1.4675	1.4983
30	1.1195	1.1307	1.1479	1.1634	1.3898	1.4209	1.4604	1.4897
35	1.1292	1.1399	1.1563	1.1710	1.3815	1.4098	1.4454	1.4715
40	1.1372	1.1475	1.1631	1.1772	1.3748	1.4008	1.4333	1.4571
50	1.1498	1.1593	1.1738	1.1868	1.3644	1.3870	1.4151	1.4354
60	1.1593	1.1682	1.1818	1.1940	1.3565	1.3768	1.4017	1.4197
70	1.1668	1.1753	1.1882	1.1996	1.3504	1.3689	1.3915	1.4077
80	1.1730	1.1811	1.1933	1.2042	1.3455	1.3625	1.3833	1.3981
90	1.1782	1.1859	1.1976	1.2080	1.3414	1.3572	1.3765	1.3903
100	1.1826	1.1900	1.2013	1.2112	1.3379	1.3528	1.3709	1.3837
150	1.1979	1.2043	1.2139	1.2223	1.3260	1.3377	1.3519	1.3619
200	1.2073	1.2130	1.2215	1.2290	1.3189	1.3289	1.3408	1.3492
300	1.2187	1.2235	1.2307	1.2369	1.3105	1.3184	1.3279	1.3344
400	1.2257	1.2299	1.2362	1.2417	1.3055	1.3122	1.3203	1.3258
600	1.2341	1.2376	1.2429	1.2474	1.2995	1.3049	1.3113	1.3158
800	1.2391	1.2422	1.2468	1.2509	1.2960	1.3006	1.3061	1.3099
1000	1.2426	1.2454	1.2496	1.2532	1.2936	1.2977	1.3025	1.3059

Note: Given n, the number of random points (sample size) in T-square sampling, enter the table with $m = 2n$; that is, with $n = 10$ points, enter this table with $m = 20$. Values of h_T below the regular alternative value indicate significant departure from a random pattern in the direction of a regular pattern, and values above the aggregate alternative indicate significant departure from a random pattern in the direction of an aggregate pattern.

the sample size equal to $2n$. At a particular level α of the test, values of h_T below the critical value shown in the table indicate a regular distribution of items, and values above the critical value indicate an aggregate distribution of items. Estimates of density are liable to be biased if a significant departure from a random distribution is indicated.

Intuitive estimates of density if items are randomly distributed are as follows:

$$\hat{D}_1 = (\text{Number of items}) / (\text{Area searched})$$

$$= n / (\text{Area of circles searched}) \tag{6.1}$$

$$= n / \Sigma(\pi x_i^2)$$

and

$$\hat{D}_2 = (\text{Number of items}) / (\text{Area of half-circles searched}) \tag{6.2}$$

$$= n / [\Sigma(\pi z_i^2 / 2)]$$

An estimator of density that is more robust to the lack of random pattern of the items is credited to Byth (1982) and takes the form

$$\hat{D}_T = n^2 / \left\{ [2\Sigma x_i][2^{\frac{1}{2}}\Sigma z_i] \right\}, \tag{6.3}$$

The reciprocal of density $1/\hat{D}_T$ is easier to handle from a mathematical point of view because it follows approximately a t distribution with $n - 1$ degrees of freedom. The standard error is given by

$$\text{SE}(1/\hat{D}_T) = \sqrt{\{8(\bar{z}^2 s_x^2 + 2\bar{x}\bar{z}s_{xz} + \bar{x}^2 s_z^2) / n\}} \tag{6.4}$$

where $\bar{x}$ is the mean of the point to the nearest-item distances, $\bar{z}$ is the mean of the item to the nearest T-square neighbor distances, s_x is the standard deviation of the point to the nearest-item distances, s_z is the standard deviation of the item to the nearest T-square neighbor distances, and s_{xz} is the covariance of the x and z distances.

An approximate 95% confidence interval on the reciprocal of density is

$$1/\hat{D}_T \pm t_{0.025,n-1} \cdot [\text{SE}(1/\hat{D}_T)], \tag{6.5}$$

where $t_{0.025,n-1}$ is the value of the t distribution with $n - 1$ degrees of freedom that is exceeded with probability 0.025. After obtaining a confidence interval for the reciprocal of the density, this can be inverted to obtain confidence limits for the density itself.

EXAMPLE 6.1 Trees in Lansing Woods

As an example of the *T*-square sampling and the corresponding density estimation using Byth's formula from Equation (6.3), consider data on the locations of 2251 trees in a 19.6-acre plot in Lansing Woods, Clinton County, Michigan, USA (Gerrard, 1969). The original plot size (924 × 924 ft) has been rescaled to unit squares. Trees are identified according to their botanical classification as hickories, maples, red oaks, white oaks, black oaks, or miscellaneous trees. For the purpose of the present example, density estimation was made only for hickories and oaks (red, white, and black oaks combined). It is worth noticing that the trees in this data set are completely mapped; thus, there are better sampling methods for density estimation than *T*-square sampling. Nevertheless, the *T*-square sampling for these data is used here for illustrative purposes, with the advantage that it is possible to compare the estimates produced by Byth's formula to the actual tree density in the rescaled plot. A systematic sample of $n = 25$ random points in a 5 × 5 grid was taken, with each random point selected from the interior of each subplot determined by the grid. Figure 6.2 shows

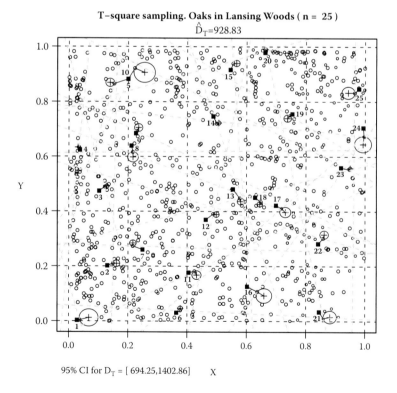

FIGURE 6.2
T-square sampling of oaks in Lansing Woods. Twenty-five points (+) in a 5 × 5 grid were randomly positioned in the unit square, and the nearest oaks to the random points (the point on each circle) were located. The squared points indicate the nearest oaks to the former oaks. Byth's estimate of the density $\hat{D}_T$ and the 95% confidence interval for D_T are also shown.

both the locations of oaks and the set of random points within the rescaled plot, Figure 6.3 shows the corresponding points for hickories, and Table 6.2 gives the results from Hines and O'Hara Hines's (1979) test of randomness, based on the critical values given in Table 6.1.

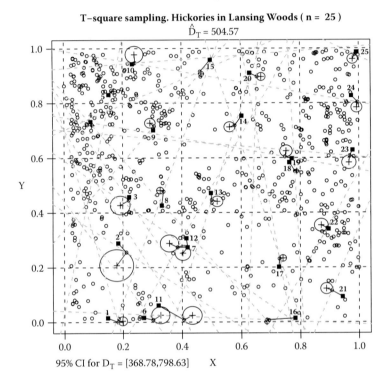

T–square sampling. Hickories in Lansing Woods (n = 25)
$\hat{D}_T = 504.57$

95% CI for D_T = [368.78,798.63]   X

FIGURE 6.3

T-square sampling of hickories in Lansing Woods. Twenty-five points (+) in a 5 × 5 grid were randomly positioned in the unit square, and the nearest hickories to the random points (the point on each circle) were located. The filled squares indicate the nearest hickories to the former nearest hickories. Byth's estimate of the density $\hat{D}_T$ and the 95% confidence interval for D_T are also shown.

TABLE 6.2

Density Estimation of Oaks and Hickories in Lansing Woods Produced by *T*-Square Sampling of $n = 25$ Random Points in a 5 × 5 Grid

Species	Hines and O'Hara Hines Statistic h_T	Byth's Estimated Density $\hat{D}_T$	95% Confidence Interval for D_T	Actual Density D_T (Unit Square)
Oaks	1.25[a]	928.83	[694.25, 1402.86]	929
Hickories	1.40[b]	504.57	[368.78, 798.63]	514

[a] Nonsignificant departure from a random pattern, $p > 0.05$.

[b] Significant departure from a random pattern in the direction of an aggregated pattern, $0.01 < p < 0.025$.

6.3 Performance of *T*-Square Sampling

Zimmermann (1991) mentions two potential unwanted effects when plotless methods are applied for the estimation of density. One is that searching for nearest neighbors using circular regions may cause them to overlap, producing dependent measurements. The other problem is that edge effects may cause problems as the distances to nearest neighbors tend to be larger near the boundary than the distances in the interior of the searching region. To cope with this second problem, several authors have suggested searching in a smaller subregion within the study area. As an alternative, Zimmermann (1991) explored censoring methods to avoid the reduction of the original study area, relying on maximum likelihood methods for the estimation of effects. He also studied the performance of some robust estimators, particularly under departures from the random location of items. Zimmermann concluded that Byth's (1982) estimator $\hat{D}_T$ appears to be efficient and robust against edge and overlap effects and departures from complete spatial randomness. However, Hall et al. (2001) argued that the need to locate a number of random points in the field reduces the attractiveness of *T*-square sampling in areas like those of forestry, where the cost of locating random points is large. These authors therefore recommended another family of plotless sampling procedures, called wandering-quarter methods (Catana, 1963), that may reduce the number of random starting points.

More tests of the performance of plotless sampling procedures were performed by Engeman et al. (1994), who concluded that the usual *T*-square estimator based on Byth's equation is in the midrange of performance. Using a simulation study, Steinke and Hennenberg (2006) compared the power of plotless density estimators (PDEs), including the *T*-square method, on plot counts of two plant populations. Tests were developed to compare the densities of two independent populations, and the performance of the tests was examined. All simulations were run for spatially random and aggregate data patterns. Steinke and Hennenberg found that for completely random data, all estimators and all tests are well behaved if the sampling intensity is the same, but for the aggregate pattern, all PDEs were negatively biased and the quadrat count estimator was unbiased.

6.4 Applications

Lamacraft et al. (1983) compared several plotless procedures seeking the most convenient formula for density estimation of three plant species in the arid rangelands of central Australia. They concluded that Byth's (1982) method was the least biased, whichever allocation procedure of points, random or semisystematic, is used. Aerts et al. (2006) applied *T*-square sampling to study the effects of pioneer shrubs on the recruitment of the tree *Olea europaea* ssp. *cuspidata* in an Ethiopian savanna. Specifically, they recorded nurse plants of seedlings using *T*-square plotless sampling.

The *T*-square method has been modified to allow the estimation of human population sizes, particularly for estimating the size of displaced human populations in emergency situations (Brown et al., 2001; Noji, 2005; Grais et al., 2006; Bostoen et al., 2007). As an example, Henderson (2009) considered occupied houses as the objects of interest, and instead of the density, the number of occupants in houses was counted. Finally, applications and alternative procedures for testing spatial patterns of objects sampled by the *T*-square method were presented by Ludwig and Reynolds (1988), Diggle (2003), Bostoen et al. (2007, Appendix 1), and ErfaniFard et al. (2008).

6.5 The Wandering-Quarter Method

The wandering-quarter method is an adaptation of the plant-centered quarter method, not requiring any assumption regarding the spatial pattern of the population studied. In fact, it provides a means of detecting regular, random, or contagious distributions. In Catana's (1963) original proposal, the data are collected from four transects positioned in the study area and grouped in two sets of two parallel transects, each set being perpendicular to the other. On each transect, the wandering-quarter method starts with an initial point-to-object distance so that the nearest individual within a 90° angle of inclusion around a compass line in the direction of the transect is selected as the starting point for measurements. This first point-to-object distance is followed by a chain of object-to-object distances so that another 90° angle of inclusion is built with the first object as a vertex and the compass line as a bisector. The distance to the nearest individual within this 90° angle is recorded; the procedure continues until one of the boundaries of the sampling area is reached or when a preestablished number of objects have been found (typically 25). See the work of Catana (1963) and Tongway and Hindley (2004) for more details. Figure 6.4 exemplifies the wandering-quarter sampling process for a single transect.

Let $d_1, d_2, \ldots, d_N$ be the N individual wandering distances between items, measured for the four transects combined, and let

$$\bar{d} = \sum_{i=1}^{N} d_i / N$$

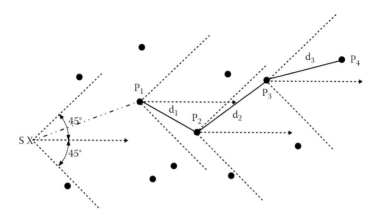

FIGURE 6.4
Wandering quarter sampling (Catana, 1963). S is a randomly selected point. P_1 is the nearest object to S within a quarter determined by the direction of the transect (indicated by the arrow), having S as the starting point for searching. The nearest object to P_1 (i.e., P_2) found in the next quarter determines the first measured distance d_1 used in the formula for density estimation. The procedure continues in a similar fashion, producing additional distances d_2, d_3, ... See text for further details.

be the mean distance between items. If items in the study region are randomly dispersed, the density is estimated as

$$\hat{D}_{wq} = A/\bar{d}^2,$$ (6.6)

where A is the unit of area, and $\bar{d}^2$ is interpreted as the mean area of all items. For a clumped spatial distribution of items, Catana (1963) developed an estimator of D_{wq} based on methods suggested by Cottam et al. (1953, 1957), that is,

$$\hat{D}_{wq} = N_{cl} \times N_{itc},$$

where N_{cl} is the number of clumps per unit area, and N_{itc} is the number of items per clump. However, here only the estimator of D_{wq} for randomly dispersed items is considered.

As noted by Diggle (2003), the wandering-quarter method is an ingenious method whose slow adoption may be because of the use of longer chains of object-to-object distances that would produce distances with increasing likelihood of boundary problems (i.e., reaching the end of the region in less than n steps either because Catana recommended $n = 25$ or because of encountering the side boundary). Another reason why Catana's method is not so popular is the lack of an easy-to-compute standard error of the estimate. Hall et al. (2001) proposed a bootstrapping procedure for a generalized version of the wandering-quarter method, but the density estimator is different from that suggested by Catana (1963).

Following the work of Diggle (2003), the randomness of the objects arranged in space can be tested using the statistic

$$I_{wq} = \sum_i \pi \hat{D}_{wq} d_i^2 / 2,$$

which is distributed as χ^2_{2N} when the spatial pattern of items is completely random (N being the total number of wandering distances).

> **EXAMPLE 6.2 Japanese Pine Point Pattern**
>
> The data in this case consist of 65 Japanese black pine (*Pinus thunbergii*) saplings in a square sampling region in a natural forest. Numata (1961) originally collected the tree locations in a square 5.7×5.7 m, but the region has been rescaled to unit squares and the point data rounded to two decimal places. Like the example given in *T*-square sampling using the Lansing Woods data, saplings are completely mapped, but the application of the wandering-quarter method and the density estimation allow a comparison between the estimate produced by Catana's formula in Equation (6.6) to the actual sapling density in the rescaled plot. The placement of transects in the unit square was simulated, where these four transects indicate the search direction of saplings and determine the quadrat (quarter) from where wandering distances are measured (see Figure 6.5). The use of Equation (6.6) in this case is completely justified because the spatial pattern of the 65 Numata's black pine saplings appears to be random (Diggle, 2003). The estimated density based on the resulting 27 wandering distances is $\hat{D}_{wq} = 63.55$ saplings per unit area.

6.6 Further Extensions and Recent Developments in Plotless Sampling Methods

Bostoen et al. (2007) provided an extension of the *T*-square method applicable to designs seeking to establish planning resource requirements or assessing health needs where sampling frames are unavailable. The procedure, basically a combination of *T*-square sampling and optimization techniques, is executed in two stages. The first one optimizes the sample size, and the second one optimizes the pathway connecting the sampling points, which entails the solution of the well-known traveling salesperson problem (Applegate et al., 2006). Coincidently, in the same article, Bostoen et al. (2007) suggest the wandering-quarter method as an alternative sampling procedure applicable to human populations.

Angles between objects can also be used for revealing spatial patterns in plotless sampling. As an example, Assunção (1994) has proposed a sampling

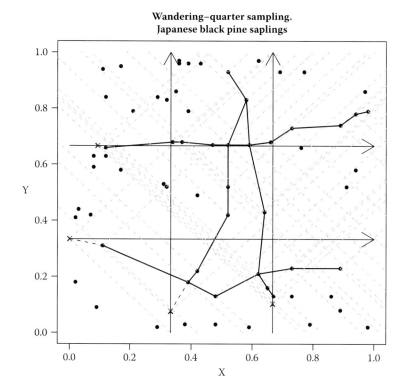

**Wandering–quarter sampling.
Japanese black pine saplings**

FIGURE 6.5
One possible placement of transects (arrows) and derived trajectories (determined by the direction of the arrows) inside quarters (delimited by the dashed lines) for the measurement of distances between Japanese black pine saplings, sampled by the wandering-quarter method. The cross × on each transect line indicates a random starting point for the search of the nearest sapling inside a quarter. The estimated density [using Equation (6.6)] is 63.55 saplings per unit area.

procedure involving the measurement of angles between the lines of sight from sampling points to their nearest two neighboring objects and the corresponding estimation of the mean angle as an index for testing spatial randomness. Furthermore, Assunção and Reis (2000) compared the performance of distance-based tests of randomness based on T^2 distances and angle-based tests and found that the Hines and O'Hara Hines procedure was the best test in terms of power. Nevertheless, Assunção's angle-based procedure has been recommended by Trifković and Yamamoto (2008) for indexing spatial patterns of trees in forest stands, but they warn that this indexing can be improved when the mean of angles is accompanied by a description of the frequency distribution of angular measurements.

  T-square sampling and the wandering-quarter method are relatively robust to the assumption of a random spatial pattern of items. However, there has recently been some interest to apply design-based approaches for density estimation in plotless sampling, without assuming a particular spatial pattern of the population of interest. Barabesi (2001) gave one of those methods, applied to the estimation of plant density, based on (random) point-to-plant distances. The procedure involves the estimation of the probability density function of these distances through a kernel density estimator.

6.7 Computational Tools for Density Estimation in Plotless Sampling

Equations entailing the estimation of densities using Byth's or any related estimator as well as their corresponding standard errors in T-square sampling are simple as the x_i and z_i distances depicted in Figure 6.1 are the only information needed. Any spreadsheet program or programming language like R can be used for calculations. In fact, an R script was prepared for producing estimates, standard errors, and plots (Figures 4.1 and 4.2) in the Lansing Woods example (see the book's companion site: https://sites.google.com/a/west-inc.com/introduction-to-ecological-sampling-supplementary-materials/home/chapter-6-r-code). As another option, the program Ecological Methodology (Exeter Software, 2009), a companion software of Krebs's (1999) book, also offers Byth's and other estimators for T-square sampling.

Regarding density estimation in wandering-quarter sampling, the basic procedure presented here can also be performed using any spreadsheet or programming language. Generally, density estimates for any of the rapid plotless sampling methods described in this chapter can be readily available to surveyors in the field, given the versatile capacities of current portable computers. However, it would be worth considering the development of special software allowing the computation of estimators for a wider variety of assumptions in plotless sampling methods (e.g., accounting for excessively clumped objects). As an example, computational tools are needed for density estimation in wandering-quarter sampling when objects are aggregated following the estimation procedure originally proposed by Catana (1963). Finally, a suitable estimate of the standard error of density in wandering-quarter sampling (e.g., by bootstrapping) is waiting to be proposed, with an algorithm implemented in computers.

7

Introduction to Mark-Recapture Sampling and Closed-Population Models

Jorge Navarro, Bryan Manly, and Roberto Barrientos-Medina

7.1 Introduction

A critical issue in many ecological studies is the estimation of population abundance, whether the population is animals or plants. This becomes particularly important in the case of animal populations because of their mobility. If we combine the interest in knowing the population size with determining which forces operate on it, then we must recognize the importance of methodological tools that help us effectively meet these tasks.

The methods commonly used to estimate the abundance of animal populations include marking, releasing, and recapturing of individuals, processes that in most cases are repeated several times. For this reason, they are called mark-recapture or capture–recapture methods. In general, these techniques allow us to estimate the size of an animal population by capturing the individuals that have been previously marked (Williams et al., 2002). Mark-recapture methods are also valuable for estimating the density of populations, that is, the number of organisms present per unit area.

The first applications of these principles were by John Graunt and Simon Laplace (both working with human populations), but C.G.J. Petersen (a fisheries biologist) made the first ecological application in 1896. Other historical applications of these methods include the estimation of the abundance of ducks made by Lincoln in 1930 and the study by Jackson in 1933 to estimate the abundance of insect populations (Williams et al., 2002; Amstrup et al., 2005).

The first part of this chapter covers some basic mark-recapture terminology useful for ecologists interested in the study of animal populations. A crucial feature that is important to identify in any study is whether the sampled population is closed or open, where closed means that the abundance does not change with time and open means that animals may enter and leave the population between sample times. Historically, the theory of closed mark-recapture methods was developed first, and it concerned less-sophisticated procedures in comparison to open-population models. Thus, this chapter focuses on mark-recapture methods applicable to closed populations, with open-population models covered in Chapter 8. Good references for the material reviewed in this chapter and Chapter 8, and for more advanced methods, are the work of Otis et al. (1978), White et al. (1982), Pollock et al. (1990), Williams et al. (2002), and Amstrup et al. (2005).

7.2 Terminology and Assumptions

Important concepts related to mark-recapture methods are as follows:

1. The sampling method. Mark-recapture methods involve taking a series of samples of the population under study, separated by one or more days or weeks, which in practice means any discrete time intervals.

2. The type of population. An important aspect of the selection of a method for the analysis of data is the type of population under study, which is assumed to be either closed or open. As noted, a closed population is one in which the population size does not change appreciably during the study period so that any gains through births and immigration or losses through mortality and emigration in the population are assumed to be minimal. In contrast with open populations, the population size changes throughout the study period.

3. The type of marking. There are three types of marks. The most natural mark is affixed, clipped, painted, and the like on each individual so that it can be identified if it is recaptured. The second type of mark is date or sample specific, indicating how many times the animal was captured throughout the study. Finally, there is the less-specific mark, which simply allows the separation between labeled and unlabeled animals and thereby provides a minimum amount of information. One aspect that should be kept in mind is the fact that the marking should not jeopardize the survival of an animal.

4. The covariates. Mark-recapture sampling may be used in a wide range of research goals involving the estimation of population parameters, where these may be ascribed to differences in individuals' characteristics (e.g., age, sex, habitat, or body weight) or particular incidents (e.g., catastrophic events like hurricanes). Thus, covariate recording might be crucial for the mark-recapture process.

Much of the development in mark-recapture methods has therefore been driven by the goal of modeling changes of population parameters in terms of one or more covariates (explanatory variables).

Along with these basic concepts, methods of capture–recapture share a number of theoretical assumptions:

1. All marks are permanent and are correctly recognized in the recaptures. This assumption applies only for the period of study.
2. There is no effect of the fact that an individual is caught, handled, and marked on one or more occasions on its later probability of capture. In other words, individual catchability is not affected by being caught.
3. The possibility that an individual dies or migrates is unaffected by its handling, and any migration is permanent and therefore indistinguishable from mortality.
4. All the animals, marked or not, have the same probability of being captured. This means that the sampling is effectively done at random, regardless of age, sex, or the physical condition of individuals.
5. All individuals, marked and not, have the same probability of dying or emigrating. That is, movements out of the population are not affected by whether individuals are or are not marked.

7.3 Closed-Population Methods

7.3.1 The Petersen–Lincoln Method

As noted above, a population is said to be closed if there is no change in the number of individuals in the population during the study. That is, there are no deaths, recruitment, emigration, or immigration in the population.

Suppose that a random sample of n_1 animals is taken from a population consisting of N animals. These animals are then marked and released back into the population. On a second capture occasion, a sample of n_2 animals is taken from the population, of which m_2 are found to be marked. Then, it might be expected that the proportion of marked animals in the second sample will be approximately the same as the proportion in the population. That is, $m_2/n_2 \approx n_1/N$, so that $N \approx n_1 n_2/m_2$. It follows that an estimate of the population size is given by

$$\hat{N} = n_1 n_2 / m_2. \tag{7.1}$$

This estimator is often called the Petersen estimate by fisheries biologists and the Lincoln index by terrestrial wildlife biologists following its independent use by Petersen (1896) and Lincoln (1930), respectively. However, the principle involved was used even earlier by Laplace (1786) to estimate the size of the human population of France and probably even before that. Here, this is called the Petersen–Lincoln estimate.

The assumptions behind Equation (7.1) are the following:

a. The population is closed so that there are no losses and gains in the time between the two samples.

b. The second sample is randomly chosen from the population.

c. No marks are lost before the second sample is taken, and all marked animals are recognized as such in the second sample.

To meet assumption a, at least approximately, the study must generally be carried out over a relatively short period of time. Actually, assumption a can be relaxed slightly. If there are losses from the population but no gains, and the losses are at the same rate for marked and unmarked animals, then $\hat{N}$ estimates the population size at the time of the first sample.

A potential problem with using Equation (7.1) is that m_2 can be zero, giving an infinite estimate for N. To overcome this, several modified estimators have been proposed, but the estimator proposed by Chapman (1951) is preferred, which is

$$\hat{N}^* = (n_1 + 1)(n_2 + 1)/(m_2 + 1) - 1. \tag{7.2}$$

This estimator is unbiased (i.e., it would give the correct average value if a study was repeated a large number of times) when $n_1 + n_2 > N$ and is approximately unbiased otherwise. Also, an estimate of the standard error of this estimator is

$$\widehat{SE}(\hat{N}^*) = \sqrt{\left[(n_1 + 1)(n_2 + 1)(n_1 - m_2)(n_2 - m_2)/\left\{(m_2 + 1)^2(m_2 + 2)\right\} \right]}. \tag{7.3}$$

One criticism of these estimators is that the equations for the standard errors are derived under the assumption that the sample sizes n_1 and n_2 are fixed before the study, so Sekar and Deming (1949) derived the estimate of the standard error of the index without this assumption:

$$\widehat{SE}(\hat{N}^*) = \left(n_1 n_2 (n_1 - m_2)(n_2 - m_2)/(m_2)^3 \right)^{\frac{1}{2}} \tag{7.4}$$

This estimator is not corrected for bias and is known to be valid only for large sample sizes, but in practice it gives results that are almost the same as those given by Equation (7.3). For this reason, Equation (7.3) should work well for small samples as well as large samples and when the sample sizes are not fixed. Also, it can be argued that the standard error conditional on the observed values of n_1 and n_2 is the one that is of interest.

EXAMPLE 7.1 The Size of a Deer Mouse Population

As an example, consider part of the results of an experiment on the effects of controlled burning on populations of the deer mouse (*Peromycus maniculatus*) as discussed by Skalski and Robson (1992, p. 126). On one unburned site, 3 days of trapping and marking produced $n_1 = 49$ marked mice. Three more days of trapping then produced $n_2 = 82$ mice, of which $m_2 = 26$ were marked. For these data, the nearly unbiased estimate from Equation (7.2) is

$$\hat{N}^* = (49 + 1)(82 + 1)/(26 + 1) - 1 = 152.7$$

with

$$\widehat{SE}(\hat{N}^*) = \sqrt{[50 \times 83 \times (49 - 26) \times (82 - 26)/(27^2 \times 28)]} = 16.2$$

This suggests that the population size was probably within the range $152.7 \pm 1.96 \times 16.2$, or 121 to 184, taking the estimate plus and minus 1.96 standard errors as an approximate 95% confidence interval with rounding to integers.

7.3.2 Sample Size Recommendations

To avoid wasting effort on a study that is not able to give a reasonable level of accuracy, it is important to plan sample sizes in advance, albeit with the recognition that planned sample sizes and obtained sample sizes may differ considerably. Robson and Regier (1964) have given recommendations for the sample sizes that will yield estimates of N using Chapman's modification of the Peterson–Lincoln index, which are within $\pi\%$ of the true value with 95% confidence. They considered values of $\pi = 50\%$, 25%, and 10% and provided sample size charts in which, for given pairs of (n_1, n_2), one is able to determine the necessary precision with 95% confidence.

After reviewing a number of different combinations of pairs (n_1, n_2), a general pattern appears. Fairly large samples are required to obtain accurate population size estimates from the Petersen–Lincoln equation. As a general rule, there should be at least 10 marked animals in the second sample, and for a large population far more than this is needed. For example, if the population size is 100 and n_1 and n_2 are to be equal, then to obtain a 95% confidence interval of about 90 to 110, it is necessary to have $n_1 = n_2 = 65$, leading to about 42 marked animals in the second sample. For a second example, suppose that N is 10,000. Then, taking $n_1 = n_2 = 800$ will give a 95% confidence interval of about 7500 to 12,500, with about 64 marked animals in the second sample. Robson and Regier's method was reviewed and exemplified by Krebs (1999).

7.3.3 Multiple Samples: The Models of Otis et al.

In the early development of mark-recapture methods, the focus was to develop more reliable estimates of the population size N for closed populations. For example, Schnabel (1938) proposed an extension of the Petersen–Lincoln estimator to allow for multiple mark-recapture events (say, over t sampling occasions), where the probability of capture is allowed to vary among sampling occasions, although for each sampling event all animals have the same probability of capture (i.e., the equal catchability assumption applies). Improvements of Schnabel's method were developed later (e.g., see Schumacher and Eschmeyer, 1943). Currently, all these related methods are known as classical mark-recapture models for closed populations, but it was often shown that the equal catchability assumption was too restrictive (e.g., see Chao and Huggins, 2005a) and could lead to severe bias in the estimators of the population size. A major advance in this area was the development of the theory for a set of eight alternative models for data from multiple samples (Otis et al., 1978; White et al., 1982). This set of models possesses a hierarchical structure of increasing complexity and capability, accounting for heterogeneity in capture probabilities. Software was developed for its use. The first computer program was CAPTURE (White et al., 1978; Rexstad and Burnham, 1992). Later, POPAN was developed, which invoked CAPTURE through a Windows interface (Arnason et al., 1998), then MARK (White and Burnham, 1999), and CARE-2 (Chao and Yang, 2003), among others.

Assume that the population under study is closed and the captures are independent events. Let P_{ij} denote the capture probability of the ith animal on the jth occasion; p_i denote the heterogeneity effect of the ith individual for $i = 1, 2, \ldots, N$; and e_j be the time effect of the jth sampling occasion for $j = 1, 2, \ldots, k$. To estimate the unknown parameter N, three sources of variations are introduced: (1) time variation in the capture probability (the t effect); (2) behavioral responses to trapping (the b effect); and (3) heterogeneity in capture probabilities for different individuals (the h effect). The combinations of one, two, or all three of these effects, plus the model with none of them, produce eight models, as follows:

M_0, equal catchability: All individuals have the same probability p of being captured on each sampling occasion ($P_{ij} = p$);

M_t, time variation: All individuals have the capture probability e_j for the jth occasion ($P_{ij} = e_j$, which is Schnabel's model);

M_h, heterogeneity: The ith individual has its own unique capture probability p_i, which remains constant for all sampling occasions ($P_{ij} = p_i$);

M_b, trap response: Every unmarked individual has the same capture probability p, which changes to c after the first capture ($P_{ij} = p$ until first capture, $P_{ij} = c$ for any recapture);

M_{bh}, heterogeneity and trap response: The ith individual has its own unique capture probability p_i before it is captured, which changes to c_i after the first capture ($P_{ij} = p_i$ until first capture and $P_{ij} = c_i$ for any recaptures, known as the generalized removal model);

M_{th}, time variation and heterogeneity: The ith individual has its own unique capture probability, which varies from sample to sample ($P_{ij} = p_i e_j$);

M_{tb}, time variation and trap response: The probability of capture is e_j for uncaptured individuals and c_j for captured individuals in the jth occasion ($P_{ij} = e_j$ until first capture, $P_{ij} = c_j$ for any recapture); and

M_{tbh}, time variation, trap response, and heterogeneity: The ith animal has its own unique capture probability, which varies with time and changes after the first capture ($P_{ij} = p_{ij}$ until first capture, $P_{ij} = c_{ij}$ for any recapture).

It is desirable to estimate the parameters of these models using the principle of maximum likelihood (ML); that is, parameters should be estimated by the values that make the probability of obtaining the observed data as large as possible. However, the usual type of mark-recapture data does not provide enough information to estimate models M_{th}, M_{tb}, and M_{tbh} by ML. This was the main problem faced by the implementation of ML estimation in CAPTURE. Nevertheless, CAPTURE still can choose between the other five models on the basis of goodness-of-fit tests. Hence, after giving a full account of the most recent estimation methods available (e.g., ML, log-linear models, moment estimators, estimating equations, jackknife, etc.), Chao and Huggins (2005b) warned that "there is no objective method to select a model from the various heterogeneous models" (page 72).

The main interest for the last 30 years has been the development of alternative estimation procedures. As an example, to reduce the bias of estimators, it is a standard practice to invoke jackknife methods (Manly, 2006). This is particularly effective for fitting model M_h. The jackknife approach is already implemented in the CAPTURE program. Jackknifing is also a suitable estimation method for model M_{bh} (Pollock and Otto, 1983); for models M_h and M_{th}, the sample coverage approach (Lee and Chao, 1994) has practical advantages as it makes it possible to summarize the heterogeneity effects in terms of the coefficient of variation (CV) of the capture probabilities, where the larger the CV the greater the degree of heterogeneity among animals (Chao and Huggins, 2005b). The program CARE-2 (Chao and Yang, 2003) implements this sample coverage approach for closed-population models. For the most complex model, M_{tbh}, CARE-2 uses a general estimation equation approach (Mukhopadhyay, 2004) by which the probability of recapture is assumed to take a multiplicative form, $c_{ij} = \varphi p_i e_j$ (where φ is the behavioral response effect), and the heterogeneity effects p_i are characterized by their mean and CV (Chao and Huggins, 2005b). The estimating equation method is applicable for all of the Otis et al. (1978) models, and CARE-2 includes these last methods and a bootstrapping approach for the estimation of population sizes and their standard errors.

Before considering model selection for mark-recapture data gathered in multiple samples, the first task is to test whether the population to be modeled is truly closed. Since the unifying work by Otis et al. (1978), a procedure (suggested by Burnham and Overton) was implemented in CAPTURE to test the null hypothesis that the individual capture probabilities are invariant over time against the alternative that some individuals (that were captured at least twice) were not present in the population at the beginning or the end of the study or both. This test is not appropriate in cases of temporary emigration for which individuals present at the start of the study then left the study for a time and returned before the end of the study. Improved tests of closure have been given by Stanley and Burnham (1999) that provided an overall closure test based on a chi-squared statistic together with component tests on whether the population experienced additions or losses during the study. In particular, these component tests compare the null hypothesis of a closed-population model M_t versus the alternative that losses or additions are governed by the open-population Jolly–Seber model that is described in Chapter 8. Computations for all these tests and for Burnham and Overton's test described previously can be performed in the CloseTest program (Stanley and Richards 2005).

Example 7.2 The Snowshoe Hares

These data, collected by Burnham and Cushwa, were analyzed in Cormack (1989) and Agresti (1994). Using the same data, Baillargeon and Rivest (2007) illustrated the application of capture–recapture analyses based on log-linear models and their implementation in the R package Rcapture. Under this approach (given that model comparison was possible by means of AIC, Akaike's information criterion), they showed that there is little evidence of a behavioral response, and the best-fitting model was heterogeneous, corresponding to a particular version of model M_{th} (called $M_{th\,Poisson2}$), producing a population size estimate of 81.1 with a standard error of 5.7 and a 95% confidence interval of 71.8 to 93.8. For this example of snowshoe hares, the data were reanalyzed using the programs CAPTURE, CloseTest, and CARE-2.

CAPTURE and CloseTest were run first to test the assumption of closure for the sampled population of hares. This indicated that the assumption of closure holds (see overall closures in Table 7.1). Additional (and more detailed) tests produced by CloseTest also suggest no additions or losses during the study (see the information on the component tests of Stanley and Burnham's Closure Text in Table 7.1). The CAPTURE program was run for model selection based on the procedures described

TABLE 7.1

Closure Tests for the Snowshoe Hares Data, Produced by the CloseTest Program
(Stanley and Richards, 2005)

	Test	Statistic	df	p
Overall Closure				
	Burnham and Overton[a]	z = −0.311	—	0.3779[b]
	Stanley and Burnham (1999)	Chi² = 3.174	8	0.9229[b]

Component Tests of Stanley and Burnham Closure Test

	Test	Chi²	df	p
Additions to population	No recruitment vs. Jolly–Seber[c]	1.421	4	0.8406[d]
	Time effect vs. no mortality	1.976	4	0.7402[d]
Losses from population	Time effect vs. no recruitment	1.754	4	0.7810[d]
	No mortality vs. Jolly–Seber[c]	1.199	4	0.8783[d]

[a] Described in Otis et al. (1978) and present in the CAPTURE **program**.
[b] Low *p* values suggest population not closed.
[c] A model for open populations (see Chapter 8).
[d] Low *p* values suggest there were additions to the population.
[e] Low *p* values suggest there were losses from the population.

by Otis et al. (1978), showing that the appropriate model is probably M_h, followed closely by models M_0 and M_{th}. CAPTURE produces an interpolated jackknife estimator (Burnham and Overton, 1978) for the population size of 87, with a standard error of 6.76 and an approximate 95% confidence interval of 78 to 105 snowshoe hares.

Applying the estimation approaches implemented in CARE-2, it is possible to include various estimators for the population size for M_h and M_{th} and gain more information about the heterogeneity in capture probabilities. The corresponding estimations are summarized in Table 7.2. Thus, the estimated coefficients of variation of the capture probabilities for all estimation methods range between 0.38 and 0.49, which provides evidence of heterogeneity. The CARE-2 program produces two standard error estimates: the asymptotic estimate and the bootstrap standard error. The asymptotic standard error for the models considered in Table 7.2 (where heterogeneity is assumed) were obtained by the delta method (Casella and Berger, 2001). In the case of the estimating equation approach (EE in the table), the asymptotic standard error is not computable for models M_h and M_{th} because of their complexity. The output also shows two types of 95% confidence intervals constructed using bootstrap standard errors. One is based on the log-transformed estimates of N, as described in Chao (1987), and the other is derived using the percentile method (Efron and Tibshirani, 1993; Manly, 2006).

Comparing the confidence intervals for models M_h and M_{th} under the sample coverage 2 estimator (SC2), it can be seen that the confidence interval for M_{th} is wider than that for M_h because more parameters are included. In capture–recapture models for closed populations, it is common to find that a simpler model has a narrow confidence

TABLE 7.2

Estimated Number of Snowshoe Hares (Cormack, 1989) for Two Heterogeneity Models M_h and M_{th} Computed by CARE-2 (Chao and Yang, 2003), Using Several Estimation Approaches

Model	Estimate	Bootstrap SE	Asymptotic SE	CV	95% CI (log-transf)	95% CI (percentile)
M_h (SC1)	89.1	9.18	8.77	0.48	(77.31, 115.71)	(78.43, 107.56)
M_h (SC2)	80.6	7.53	7.33	0.38	(72.28, 105.23)	(71.82, 96.97)
M_h (JK1)	88.8	6.01	6.18	—	(79.97, 104.27)	(82.17, 96.33)
M_h (JK2)	93.8	9.14	9.40	—	(81.12, 118.59)	(80.17, 109.83)
M_h (IntJK)	88.8	8.47	6.18	—	(77.68, 112.86)	(81.33, 107.48)
M_h (EE)	85.3	7.34	—	0.48	(75.82, 106.43)	(75.64, 98.36)
M_{th} (SC1)	89.4	8.88	8.85	0.49	(77.82, 114.75)	(79.24, 106.39)
M_{th} (SC2)	80.9	7.91	7.41	0.39	(72.28, 107.04)	(71.73, 98.89)
M_{th} (EE)	84.7	7.11	—	0.49	(75.53, 105.20)	(74.36, 97.09)

Note: SE, standard error; CV, coefficient of variation; CI, confidence interval; log-transf, log-transformed interval; SC1, SC2, sample coverage approaches 1 and 2, respectively (Lee and Chao, 1994); JK1, JK2, jackknife approaches 1 and 2, respectively (Burnham and Overton, 1978); IntJK, interpolated jackknife (Burnham and Overton, 1978); EE, estimating equations approach (Chao et al., 2001).

interval because "more general models produce wide intervals with a better coverage probability" (Chao and Yang, 2003, page 10). Selecting the approach based on estimating equations (Chao et al., 2001), it is estimated that there are 84.7 or about 85 hares in the population with an estimated bootstrap standard error of 7.11 and an associated 95% confidence interval of (74, 97) hares, based on the percentile method.

7.3.4 Huggins's Models

One of the major improvements in the mark-recapture analysis of closed populations is the use of a regression-type parameterization of covariates for estimating capture probabilities, with generalized linear modeling. Because capture probabilities must be within the range 0 to 1, it is sensible to build this constraint into a model. One way that this can be done involves replacing P_{ij} with a logistic function. For example, in the time-varying model M_t, the probabilities of capture and recapture are given by $P_{ij} = e_j$, and these can be modeled as

$$P_{ij} = e_j = \exp(a + c_j)/\{1 + \exp(a + c_j)\}.$$

This equation is equivalent to $\log\{e_j/(1 - e_j)\} = a + c_j$, which is a particular case of what is called the logit transformation; in generalized linear modeling terminology, this transformation is the logit link function. The logit is the most commonly used link function for modeling capture and recapture probabilities and is, for example, the default in the R package mra (McDonald, 2012), but other links are also possible. The use of generalized linear models for the probabilities of capture and recapture has the advantage of making it easy to incorporate the effects of covariates into the model, as it is done with logistic regression. For example, assuming no individual heterogeneity, if R_j is a measure of the effort put into recapturing animals in sample j, then it may be considered appropriate to model the probabilities of capture by

$$e_j = \exp(r_0 + r_1 R_j)/\{1 + \exp(r_0 + r_1 R_j)\}.$$

Using logistic functions in this way allows a good deal of flexibility in the modeling process. Several covariates can be used, and capture probabilities can be allowed to vary with time, the age of animals, an animal's group, and so on. In all cases, the likelihood is maximized with respect to the parameters in the logistic models (e.g., the c_j's or the r_j's in the two models shown previously). Once these parameters have been estimated, the logistic functions can be used to determine estimated capture probabilities for any animal, as proposed for closed populations by Huggins (1989, 1991) and Ahlo (1990).

Huggins's method involves the modeling of the capture probabilities P_{ij}, assuming that measured covariates can account for unequal catchability on different capture occasions and that the covariate values for uncaptured individuals are not known (Huggins, 1991). This assumption then requires estimation of parameters by maximization of the conditional likelihood based on the data for captured individuals. As a consequence, the estimation of N is more complicated for models including individual covariates (e.g., age, sex, body weight, etc.) or capture occasion covariates (e.g., the precipitation on each capture occasion, the effort on each capture occasion, etc.) as described by Chao and Yang (2003) in their manual for the program CARE-2. The general logistic model for P_{ij} used in this program is

$$\text{logit}(P_{ij}) = \log(P_{ij}/(1 - P_{ij})) = a + c_j + vY_{ij} + \boldsymbol{\beta}' \mathbf{W}_i + \mathbf{r}' \mathbf{R}_j, \tag{7.5}$$

where a is the intercept; $c_1, \ldots, c_{k-1}$ account for the capture occasional or time effect, with $c_k = 0$; $\mathbf{W}_i'$ is a vector of the s covariates measured on each individual i; $\boldsymbol{\beta}'$ is the vector of effects for these covariates; $\mathbf{R}_j$ is a vector of g covariates for the jth capture occasion; and $\mathbf{r}'$ denotes the vector of effect for these covariates. The Y_{ij} are dummy variables indicating whether the ith animal has been captured at least once before the jth occasion ($Y_{ij} = 1$) or not captured before that occasion ($Y_{ij} = 0$), and v is the corresponding effect of this behavioral response.

Huggins's method can be used with the eight Otis et al. models described previously. Thus, under the Huggins method, model M_0 is given by

$$\text{logit}(P_{ij}) = a,$$

and the time-varying M_t is given by

$$\text{logit}(P_{ij}) = a + c_j,$$

with $c_k = 0$. For the logit models corresponding to the remaining six models, see the work of Huggins (1991) and Chao and Huggins (2005b). Under Huggins's method, closed-population models have a much broader range than the Otis et al. (1978) models, as any individual or occasional covariate accounting for individual heterogeneity and time effects can be included in the models covered by Equation (7.5). In fact, it is customary in the mark-recapture literature of closed populations to denote by $M_0^*, M_t^*, \ldots, M_{tbh}^*$ the set of models involving the effects accounted for in the Otis et al. models, together with possibly one or more individual or occasional covariates.

As information criteria indices are available for the generalized linear models fitted, model selection is practicable. Finally, supposing that a total of n individuals were captured, the population size is estimated by the Horvitz–Thompson estimator,

$$\hat{N}_{HT} = \sum_{i=1}^{n} \{1 - \prod_{j=1}^{k} (1 - \hat{P}_{ij})\}^{-1} \tag{7.6}$$

and the corresponding standard error of $\hat{N}_{HT}$ is approximated by the asymptotic equations given by Huggins (1989, 1991).

Example 7.3 A Hypothetical Study

A hypothetical data set was built based on an example code included in the documentation of the R package mra (McDonald, 2012). The resulting encounter history for 29 animals captured or recaptured on five sampling occasions is summarized in Table 7.3, while Table 7.4 shows part

TABLE 7.3

Summary Statistics of the Hypothetical Encounter History of 29 Animals for $t = 5$ Sampling Occasions

j	u_j	m_j	n_j	M_j	f_j
1	14	0	14	0	3
2	7	5	12	14	12
3	4	11	15	21	9
4	2	14	16	25	5
5	2	15	17	27	0
				$M_6 = 29$	

Note: u_j, number of first captures at time j, $j = 1, \ldots, 5$; m_j, number of recaptures at time j; n_j, $u_j + m_j$ = number of animals captured at time j; M_j, $u_1 + u_2 + \ldots + u_{j-1}$ = number of marked animals just before the jth occasion; M_6, total number of distinct animals recorded in the study; f_j, number of animals captured exactly j times throughout the whole study.

TABLE 7.4

Estimated Size of the Hypothetical Population Summarized in Table 7.3 Under Different Logistic Models Fitted by Huggins's Method

Model	Estimate	SE	AIC	95% CI
M_0^*	30.00	1.10	201.15	(29.18, 34.72)
M_t^*	29.96	1.09	207.16	(29.16, 34.73)
M_b^*	30.82	2.08	202.46	(29.30, 39.89)
M_h^*	30.00	1.08	203.14	(29.18, 34.63)
M_{tb}^*	30.14	1.96	209.14	(29.12, 40.37)
M_{th}^*	29.96	1.09	209.15	(29.16, 34.73)
M_{bh}^*	30.82	2.08	204.45	(29.30, 39.19)
M_{tbh}^*	30.15	1.89	211.13	(29.12, 39.85)

Note: With the exception of M_0^*, M_t^*, M_b^*, and M_{tb}^*, fitted models included the categorical covariate "sex." SE, standard error; AIC, Akaike's information criterion; CI, confidence interval.

of the output of the program CARE-2 (Chao and Yang, 2003). Sex was also recorded as a dummy variable for each animal and included in the analysis as an individual covariate.

The application of the Huggins method to these data leads to different estimates of N for the different logistic models, as shown in Table 7.4. The smallest AIC was attained for the simplest model M_0^*, indicating that no temporal, behavioral, or individual heterogeneity effects were present in the capture histories. The population size estimate under model M_0^*, obtained from Equation (7.6), is 30 with a 95% confidence interval of (29.18, 34.72).

7.4 Recent Advances for Closed-Population Models

New methodologies for closed-population models were discussed by Chao and Huggins (2005a, 2005b), with detailed descriptions of methods illustrated using the two examples described previously in this chapter. This includes estimation involving bootstrapping, improved confidence intervals, sample coverage methods, estimating equation methods, and generalized linear models. Table 4.1 in the work of Chao and Huggins (2005b) gives a complete list of more recent methods, including Bayesian methods, parametric modeling of heterogeneity, latent class models, mixture models, and the use of nonparametric ML. The introduction of covariates to account for variation in probabilities of capture has been an important avenue of research based on conditional likelihood theory and generalized linear models (Huggins and Hwang, 2011).

Appendix A in Amstrup et al. (2005) provides a list of software programs for mark-recapture analysis with closed-population models, namely, CARE-2,

CAPTURE, MARK, POPAN-5, and NOREMARK (White, 1996). The classic closed-population models of Otis et al. (1978) and some updated estimators (e.g., jackknife for models M_h or M_{bh}) could be run using the original program CAPTURE in PC-DOS. Later, for PC-DOS and the earliest versions of Windows, an interactive interface for CAPTURE, called 2CAPTURE, was created (Rexstad and Burnham, 1992), which can still be run using emulators of old Windows versions. Afterward, MRI, another interactive version of CAPTURE, was created by Landcare Research (New Zealand) and built in as a module of POPAN-5. MRI also includes an interactive access to JOLLY, a program for the analysis of open populations based on the Jolly–Seber model (described in Chapter 8). The last version of MRI was Windows NT executable; thus, it can only be run using Windows emulators. At the Patuxent Wildlife Research Center (U.S. Geological Survey) website (http://www.mbr-pwrc.usgs.gov/software/capture.html), there is also a facility for running CAPTURE statements. The program MARK also includes a menu allowing access to all the CAPTURE models, including the standard Otis et al. (1978) models, Huggins's models, and other approaches for heterogeneity. However, MARK is not easy to use for a novice, and anyone interested in using it is advised to take part in a workshop that only addresses this program.

Finally, the program DENSITY is suitable for the analysis of spatially explicit capture–recapture data (Efford, 2012). This incorporates various closed-population models for the estimation of population sizes and densities of animals detected or captured from arrays of traps. See also the work of Efford et al. (2004) and Efford and Fewster (2013).

8

Open-Population Mark-Recapture Models

Bryan Manly, Jorge Navarro, and Trent McDonald

8.1 Introduction

Many of the developments in mark-recapture methodology have been designed for open animal populations, with new animals entering through births and immigration and animals leaving through deaths and emigration. Studies usually involve sampling the population several times, with animals suitably marked when they are first captured so that they can be recognized when they are recaptured and a record obtained of the captures and recaptures of individual animals. Studies of open populations often cover extended time periods, and the population changes that occur are of great interest to ecologists and managers. Early methods for analyzing data from open populations were proposed by Jackson (1939, 1940, 1944, 1948), Fisher and Ford (1947), Leslie and Chitty (1951), and Leslie et al. (1953).

8.2 The Jolly–Seber Model

An important contribution to the literature on mark-recapture methods for open populations was the Jolly–Seber (JS) model (Jolly, 1965; Seber, 1965), which is at the heart of Pollock's (1991) diagram showing the relationship between different open-population models (Figure 8.1). This model provides explicit parameter estimators with variances on the assumption that (1) every animal present in the population has the same probability p_j of being captured in the jth sample taken at time t_j; (2) every marked animal present in the population immediately after the jth sample has the same probability ϕ_j of surviving until the $(j + 1)$th sample is taken; (3) marks are not lost or not

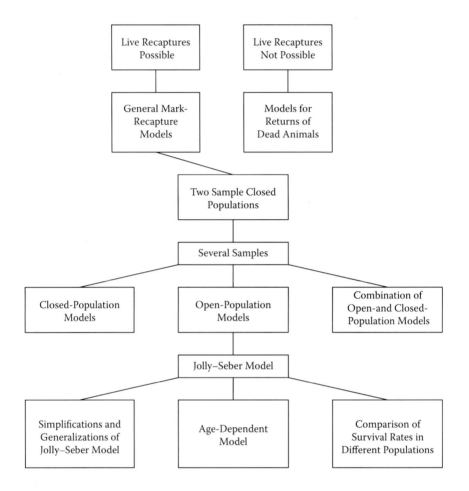

FIGURE 8.1
Relationships between models for mark-recapture data. (Based on figures provided by Pollock, K.H. *Journal of the American Statistical Association* 86: 225–238, 1991.)

recognized; and (4) all samples are effectively instantaneous and releases are made immediately after sampling.

The JS estimators were derived using the principle of maximum likelihood. However, they can also be justified intuitively in the manner described next. First, consider the estimation of the number M_j of marked animals in the population just before the jth sample. This consists of the m_j marked animals seen in the jth sample plus those that were in the population but not captured. To estimate the latter, it can be argued that the probability of an animal in the population at the time of the jth sample being recaptured in a later sample can be estimated by $z_j/(M_j - m_j)$, where z_j is the number of animals seen before and after the jth sample but not in the jth sample, and also by r_j/R_j, where R_j is the number of animals released from the jth sample of which r_j are recaptured later. Hence, it is expected that $z_j/(M_j - m_j) \approx r_j/R_j$, so that $M_j \approx m_j + R_j z_j/r_j$, suggesting the estimator

$$\hat{M}_j = m_j + R_j z_j/r_j \tag{8.1}$$

for M_j, which can be evaluated for $j = 2, 3, \ldots, k - 1$ because for these values of j all the values on the right-hand side can be obtained from the data.

Once M_j is estimated, the population size N_j at time j can be estimated because the proportion of marked animals in the jth sample should be approximately equal to the proportion in the population, that is, $m_j/n_j \approx M_j/N_j$. Thus, an estimator of N_j is

$$\hat{N}_j = n_j \hat{M}_j / m_j, \tag{8.2}$$

which can be evaluated for $j = 2, 3, \ldots, k - 1$.

To estimate ϕ_j, the survival rate over the period j to $j + 1$, it can be noted that there are $M_j + R_j - m_j$ marked animals in the population just after the releases from the ith sample, and that M_{j+1} of these are still alive at time $j + 1$. Thus, $\phi_j \approx M_{j+1}/(M_j + R_j - m_j)$, suggesting the survival estimator

$$\hat{\phi}_j = \hat{M}_{j+1} / (\hat{M}_j + R_j - m_j). \tag{8.3}$$

This can be evaluated for $j = 1, 2, \ldots, k - 2$, taking $\hat{M}_1 = 0$ because there are no marked animals in the population just before the time of the first sample.

Finally, B_j, the number of new entries to the population between samples j and $j + 1$ that are still alive at time $j + 1$, can be estimated by setting it equal to the estimated difference between the population size at time $j + 1$ and the expected number of survivors from time j, that is,

$$\hat{B}_j = \hat{N}_{j+1} - \hat{\phi}_j(\hat{N}_j + R_j - m_j), \tag{8.4}$$

which can be evaluated for $j = 2, 3, \ldots, k - 2$.

One important feature of Equations (8.1) to (8.4) is that it is not necessary for R_j, the number of animals released from the jth sample, to equal the sample size n_j. This means that damaged animals do not need to be released, and extra animals can be added to the population if desired. Of course, if the number released does not equal the number captured, then this will change population sizes from what they would otherwise have been.

The JS estimators are biased for small samples, and infinite estimates are possible because of divisions by zero. To overcome this problem, approximately unbiased estimators can be obtained by replacing $\hat{M}_j$ and $\hat{N}_j$ by

$$\hat{M}_j^* = m_j + (R_j + 1)z_j / (r_j + 1), \tag{8.5}$$

and

$$\hat{N}_j^* = (n_j + 1)\hat{M}_j^* / (m_j + 1) \tag{8.6}$$

respectively (Seber, 1982, p. 304), leaving the other estimation equations unchanged, except that the new estimators are used for M_j and N_j.

Variance equations were provided by Jolly (1965) and Seber (1965). In using these, it is necessary to distinguish between sampling errors (the difference between estimated parameters and the parameters occurring in the realized population) and stochastic errors (the difference between the realized parameters and their mean values from hypothetical repeated generations of the population). For the population size, it seems to be generally accepted that only sampling errors are relevant, so that an estimate of the variance of $\hat{N}_j$ given N_j is needed. This is approximately

$$\widehat{\mathrm{Var}}(N_j^* | \hat{N}_j) \approx \hat{N}_j[\hat{N}_j - n_j][(\hat{M}_j - m_j + R_j)(1/r_j - 1/R_j)/\hat{M}_j + (\hat{N}_j - \hat{M}_j)/(\hat{N}_j m_j)]. \tag{8.7}$$

On the other hand, it is usually considered that survival probability is being estimated rather than the proportion surviving in the population. Therefore, the variance of ϕ_j is taken to include stochastic errors. It can then be approximated by

$$\widehat{\mathrm{Var}}(\hat{\phi}_j) \approx \hat{\phi}_j^2[\{(\hat{M}_{j+1} - m_{j+1})(\hat{M}_{j+1} - m_{j+1} + R_{j+1})/\hat{M}_{j+1}^2\}(1/r_{j+1} - 1/R_{j+1}) \tag{8.8}$$

$$+ (\hat{M}_j - m_j)/(\hat{M}_j - m_j + R_j)(1/r_j - 1/R_j) + \hat{\phi}_j^2(1 - \hat{\phi}_j)/\hat{M}_{j+1}$$

Variance equations for $\hat{M}_j$ and $\hat{B}_j$ can be found in the original works of Jolly (1965) and Seber (1965) and in reviews such as those of Seber (1982, 1986, 1992), Pollock et al. (1990), and Schwarz and Seber (1999). Covariance equations are also available from the same sources.

Before the JS model was developed, there was a crucial article by Cormack (1964). He derived one component of the likelihood used by Jolly and Seber in their more general model. To recognize Cormack's contribution to the development of these models, the term *Cormack–Jolly–Seber* (CJS) model is often used when referring to the marked animal component of the likelihood function, which allows the estimation of survival and capture probabilities. Jolly (1965) and Seber (1965) used different likelihoods but came up with the same estimators.

The JS likelihood can be viewed as three conditionally independent components with the overall likelihood the product of these. The overall likelihood is then $L_1 \times L_2 \times L_3$, where L_1 is a product binomial likelihood that relates the unmarked population and sample sizes at each time to the capture probabilities, L_2 only contains parameters for probabilities of being lost on capture, and L_3 is the component that contains all the recapture information conditional on the numbers of marked animals released at different times and the parameters for capture and survival probabilities.

L_3 is the likelihood that was originally derived by Cormack (1964). One way to view estimation in the full JS model is that capture and survival probabilities are estimated from L_3 only. This is then what is called the CJS model. Once capture probabilities are estimated, the size of the population at the time of sample j can be estimated from $n_j \approx N_j p_j$, leading to

$$\hat{N}_j = n_j / \hat{p}_j, \tag{8.9}$$

which turns out to be the same as the JS estimator.

Accurate estimates can only be obtained from the JS model if the data are extensive because so many parameters are involved. For this reason, alternative models with fewer parameters, such as those with a constant survival rate per unit time, are of interest. Generalizations of the JS model allowing for trap response, the analysis of data from several cohorts of animals, and age-dependent survival rates are also available.

Before and after the JS model was introduced, various special cases and generalizations were developed and included in the computer program JOLLY. The deaths-only model is useful if there is no immigration and recruitment is not occurring. The births-only model is useful if there is no mortality or emigration. Versions of the model with constant survival or capture probabilities are also available in the JOLLY program. Some generalizations of the JS model that allow for temporary effects on survival and capture rates have also been developed and are available in the programs JOLLY or MARK. Although the JS model is intuitive and can be applied in some situations, most real-world multiple-occasion surveys will want to apply more sophisticated models described later in this chapter.

EXAMPLE 8.1 Estimation for a Moth Population

To illustrate the use of mark-recapture methods, consider Manly and Parr's (1968) data obtained from sampling an isolated population of the burnet moth *Zygaena filipendulae* (Table 8.1). This was an open population so that any analysis should allow for ingress and egress. The JS estimation equations are therefore applied to the data. Because the study involved five samples taken on 19, 20, 21, 22, and 24 July 1968, this permits the estimation of the population size on 20, 21, and 22 July; survival rates for the periods 19–20, 20–21, and 21–22 July; and the number of new entries to the population for the periods 20–21 and 21–22 July.

From these data, the values m_j, R_j, z_j, and r_j needed for Equations (8.3) to (8.6) can be obtained, and hence estimates of population parameters with their estimated standard errors are obtained as shown in Tables 8.2 and 8.3. The 95% confidence intervals shown in this table are not simply Estimate ± 1.96 (Standard Error), as is often used. Rather, they were obtained using transformations that have been derived to obtain improved limits as described by Manly (1984). The calculations were done using the computer program JS that is described in the Supplementary Material to this chapter, which can be found at https://sites.google.com/a/west-inc.com/introduction-to-ecological-sampling-supplementary-materials/. R code demonstrating calculation of the JS maximum likelihood estimates using the 'mra' package is also available at that site.

TABLE 8.1

Manly and Parr's Data from Sampling a Population of the Burnet Moth in Dale, Pembrokeshire, Wales, in July 1968

Capture Pattern	Number of Moths	July				
		19	20	21	22	24
1	1	1	1	1	1	1
2	1	1	1	1	1	0
3	9	1	1	1	0	0
4	4	1	1	0	0	1
5	10	1	1	0	0	0
6	5	1	0	1	1	1
7	2	1	0	1	0	1
8	1	1	0	1	0	0
9	2	1	0	0	1	1
10	1	1	0	0	0	1
11	1	0	1	1	1	0
12	4	0	1	1	0	1
13	4	0	1	1	0	0
14	1	0	1	0	1	1
15	2	0	1	0	1	0
16	2	0	1	0	0	1
17	1	0	0	1	1	1
18	3	0	0	1	1	0
19	7	0	0	1	0	1
20	5	0	0	0	1	1
21	21	1	0	0	0	0
22	12	0	1	0	0	0
23	13	0	0	1	0	0
24	9	0	0	0	1	0
25	19	0	0	0	0	1

Note: There were 25 distinct capture patterns obtained, as indicated by the five right-most columns of the table, where 0 indicates no capture and 1 indicates capture; the second column shows how many moths displayed these patterns. Thus, one moth was seen in all five samples, one moth was seen in the first four samples only, nine moths were seen in the first three samples only, and so on.

TABLE 8.2

Sample Statistics Obtained for the Jolly–Seber Equations for the Burnet Moth Data in Table 8.1

Date	n_i	z_i	r_i	m_i
		Sample Statistics		
19	57	0	36	0
20	51	11	29	25
21	52	12	25	28
22	31	20	15	17
24	54	0	0	35

TABLE 8.3

Estimates of Population Parameters Based on the Sample Statistics Shown in Table 8.2

Date	N	SE	Confidence Limits	ϕ	SE	Confidence Limits	B	SE
19	—	—	—	0.77	0.10	0.60 to 0.98	—	
20	88.1	12.0	72 to 126	0.75	0.11	0.56 to 1.01	29.9	11.2
21	95.9	14.0	77 to 141	0.74	0.15	0.50 to 1.12	29.9	16.1
22	101.3	22.2	72 to 176	—	—	—	—	—

Note: The confidence limits are approximate 95% limits.

8.3 The Manly–Parr Method

The data from the last example were originally used to illustrate an alternative to the JS method for estimation that does not require the assumption that the probability of survival is the same for all animals (Manly and Parr, 1968). It does, however, require all animals to have the same probability of capture.

The basis of the method is to select a group of animals that are known to have been alive at the time when the jth sample was taken because they were seen both before and after that time. Suppose that there are C_j of these individuals, of which c_j are captured in sample j, and that the total number of individuals captured at time j is n_j. Then, p_j, the probability of capture at time j, can be estimated by

$$\hat{p}_j = c_j / C_j. \tag{8.10}$$

This then makes it possible to estimate N_j, the population size at time j; ϕ_j, the fraction of the animals surviving the time between sample j and sample $j + 1$; and B_j, the number of new entries to the population between the time of sample j and the time of sample $j + 1$ that are still alive at the time of sample $j + 1$.

Estimation of the population size is possible because it is expected that $n_j \approx N_j p_j$. Thus, the estimator given in Equation (8.9) is applicable ($\hat{N}_j = n_j / \hat{p}_j$), with $\hat{p}_j$ given by Equation (8.10). This then has a sample variance that can be estimated by

$$\widehat{\text{Var}}(\hat{N}_j) = \hat{N}_j (C_j - c_j)(n_j - c_j) / c_j^2 \tag{8.11}$$

(Manly, 1969).

Estimation of the survival rate is based on $m_{j,j+1}$, the number of animals seen in both sample j and sample $j + 1$. It is expected that $m_{j,j+1} \approx n_j \phi_j p_{j+1}$, suggesting the estimator

$$\hat{\phi}_j = m_{j,j+1} / (n_j \hat{p}_j) . \tag{8.12}$$

No variance equation has yet been produced for this equation.

Having estimated N_j and ϕ_j, there is the obvious estimator

$$\hat{B}_j = \hat{N}_{j+1} - \hat{\phi}_j \hat{N}_j \tag{8.13}$$

for B_j.

The use of the idea behind the Manly–Parr method has been extended to when animals have known ages (Manly et al., 2003). Given the validity of certain assumptions, this leads to a relatively simple analysis of data using logistic regression to allow for the effects of covariates on capture probabilities. Survival rates do not have to be modeled with this approach.

8.4 Recoveries of Dead Animals

It is sometimes possible to obtain information about where and when some marked animals died by using a suitable marking or tagging system. This information can then be employed to estimate survival rates and to study the movement of the animals. The main application of studies of this nature has been with banded birds, with schemes set up to encourage anyone finding one of the bands (on or off a bird) to return it to an address on the bands.

A major contribution to the analysis of the data obtained from the recoveries of dead animals is provided by the handbook of Brownie et al. (1985), in which a series of models with different assumptions about survival and recapture probabilities is described. These models allow for animals to be banded as adults only, as young and adults, or as young, subadults, and

TABLE 8.4

Data Collected from a Bird-Banding Experiment with Releases for k Years and Recoveries of Dead Animals for l Years

Year of Release	Number Released	Recoveries of Dead Animals in Year					Total Recovered
		1	**2**	**3**	**...**	l	
1	N_1	a_{11}	a_{12}	a_{13}	...	a_{1l}	R_1
2	N_2		a_{22}	a_{23}	...	a_{2l}	R_2
3	N_3			a_{33}	...	a_{3l}	R_3
⋮	⋮				⋱	⋮	⋮
k	N_k					a_{kl}	R_k
C_j, year total		C_1	C_2	C_3	...	C_l	

adults. Survival and recovery results can either be constant or vary with the age class or the year. Estimates from the models cannot in general be calculated explicitly. Instead, the numerical solution of equations is required, using appropriate computer software.

Although it is not realistic to review the many models that have been proposed for the recovery of dead animals, it is instructive to consider one model in some detail to understand the nature of the modeling process. The model chosen is the first one discussed by Brownie et al. (1985, p. 15).

Suppose that batches of birds are banded in years $1, 2, \ldots, k$, and records of recoveries are available for years $2, 3, \ldots, k, k + 1, \ldots, l$. Then, the basic data can be displayed as shown in Table 8.4. Thus, N_i birds are released in year i, a_{ij} of these are recovered dead (or their bands are recovered) in year j, and $N_i - R_i$ are never recovered. If the survival rate for all birds in year j is S_j, and the probability of recovering a bird that dies in that year (or its band) is f_j, then the probabilities shown in Table 8.5 apply. For example, the probability that one of the birds released in year 2 is recovered dead in year 4 is $S_2 S_3 f_4$.

TABLE 8.5

Probabilities Associated with the Recovery Histories Shown in Table 8.4

Year of Release	Number Released	Recoveries of Dead Animals in Year					Probability Recovered
		1	**2**	**3**	**...**	l	
1	N_1	f_1	$S_1 f_2$	$S_1 S_2 f_3$	...	$S_1 \ldots S_{l-1} f_l$	θ_1
2	N_2		f_2	$S_2 f_3$	...	$S_2 \ldots S_{l-1} f_l$	θ_2
3	N_3			f_3	...	$S_3 \ldots S_{l-1} f_l$	θ_3
⋮	⋮				⋱	⋮	⋮
k	N_k					$S_k \ldots S_{l-1} f_l$	θ_k

Note: The recovery probability θ_i is the sum of the recovery probabilities for years i to l given in row i of the table, $1 \le i \le k$. For example θ_3 is the sum of the recovery probabilities for years 3 to l as shown in row three of the table.

For this particular model, it is possible to obtain explicit estimates for most of the unknown parameters based on the principle of maximum likelihood. The estimator of the recovery rate in year j is

$$\hat{f}_j = (R_j C_j) / (N_j T_j) \tag{8.14}$$

for $j = 1, 2, \ldots, k$, where R_j is the total number of birds recaptured from the N_j released in year j, and C_j is the total number of birds recovered dead in year j (Table 8.4). The definition of T_j is more complicated, and values must be calculated from the following relationships:

$$T_j = \begin{cases} R_j, & j = 1 \\ R_j + T_{j-1} - C_{j-1}, & j = 2, 3, \ldots, k \\ T_{j-1} - C_{j-1}, & j = k+1, k+2, \ldots, l \end{cases}$$

The estimator of the yearly survival rate S_j is

$$\hat{S}_j = (R_j/N_j)(1 - C_j/T_j)(N_{j+1} + 1)/(R_{j+1} + 1), \tag{8.15}$$

$j = 1, 2, \ldots, k - 1$, where the addition of 1 to N_{j+1} and R_{j+1} has been made to reduce bias. The maximum likelihood estimators do not include these additions.

Approximations for the variances of the estimator are provided by the equations

$$\widehat{\text{Var}}(\hat{f}_j) \approx \hat{f}_j^2 (1/R_j - 1/N_j + 1/C_j - 1/T_j) \tag{8.16}$$

and

$$\widehat{\text{Var}}(\hat{S}_j) \approx \hat{S}_j^2 \{1/R_j - 1/N_j + 1/R_{j+1} - 1/N_{j+1} + 1/(T_{j+1} - R_{j+1}) - 1/T_j\}. \tag{8.17}$$

Standard errors are estimated by the square roots of the estimated variances, and approximate $100(1 - \alpha)\%$ confidence intervals are given in the usual way by estimates plus and minus multiples of the standard errors.

Variations on this model discussed by Brownie et al. (1985) include those for which either the survival rates S_j or the recovery rates f_j (or both) are constant with time, with the fit of different models to the observed data assessed by appropriate statistical tests.

EXAMPLE 8.2 Recoveries of Male Wood Ducks

Brownie et al. (1985, p. 14) give an example of banding and recovery data for male wood ducks (*Aix sponsa*) banded before the hunting seasons in 1964 to 1966 in a midwestern state in the United States, with recoveries of dead ducks recorded for 1964 to 1968. The recovery data are reproduced here in Table 8.6 together with the derived statistics R_j, C_j, and T_j.
From Equation (8.14), it is found that

$$\hat{f}_1 = (265 \times 127)/(1603 \times 265) = 0.0792,$$

so that it is estimated that bands were recovered from 7.9% of the birds that died in 1964. From Equation (8.16), the estimated variance is

$$\widehat{\text{Var}}(\hat{f}_1) = 0.0792^2(1/265 - 1/1603 + 1/127 - 1/265) = 0.00004548$$

giving an estimated standard error of $\sqrt{0.00004548} = 0.0067$. An approximate 95% confidence interval for the true recovery rate is therefore $0.0792 \pm 1.96 \times 0.0067$ or 0.066 to 0.092 (6.6% to 9.2%). Similar calculations for the other recovery rates give $\hat{f}_2 = 0.0401$ with standard error and 95% confidence limits of 0.032 to 0.048, and $\hat{f}_3 = 0.0688$ with standard error 0.0061 and 95% confidence limits of 0.057 to 0.081. From Equation (8.15), it is found that

$$\hat{S}_1 = (265/1603)\,(1 - 127/265)\,(1596/211) = 0.651,$$

so that it is estimated that 65.1% of ducks alive at the start of 1964 survived that year. From Equation (8.17), the estimated variance is

$$\widehat{\text{Var}}(\hat{S}_j) = 0.651^2\{1/265 - 1/1603 + 1/210 - 1/1595 + \\ 1/(348 - 210) - 1/265 = 0.004556,$$

giving an estimated standard error of $\sqrt{0.004556} = 0.0675$. An approximate 95% confidence interval for the population survival rate is then

TABLE 8.6

Banding and Recovery Data for Male Wood Ducks Banded before the Hunting Seasons in a Midwestern State in the United States

Year of Release	Number Released	Recoveries of Dead Ducks in					R, Total Recovered
		1964	1965	1966	1967	1968	
1964	1603	127	44	37	40	17	265
1965	1595		62	76	44	28	210
1966	1157			82	61	24	167
C, year total		127	106	195	145	69	
T (see text)		265	348	409	214	69	

0.651 ± 1.96 × 0.0675 or 0.519 to 0.784. It is also possible to estimate the 1965 survival rate. For this, $\hat{S}_2 = 0.631$, with standard error 0.0647, and the approximate 95% confidence interval is 0.504 to 0.758.

8.5 Estimation Using Radio-Tagged Individuals

Advances in technology in recent years have made it possible to monitor the movement of even small animals by fitting them with radio-tags. This allows the use of mark-recapture methods in which the animals fitted with transmitters are thought of as marked but there is a substantial advantage over more conventional methods of marking because it becomes possible to know which marked animals are present in a study region without needing to take into account survival and migration.

Essentially, the radio-tagged animals that are known to be in the study region when a sample of animals is taken serve as the marked animals for the estimation of population size. Formally, either the Petersen–Lincoln estimator $\hat{N} = n_1 n_2 / m_2$ or, better still, the bias-corrected estimator

$$\hat{N}^* = (n_1 + 1)(n_2 + 1)/(m_2 + 1) - 1, \tag{8.18}$$

is used where n_1 is the number of marked (radio-tagged) animals available for capture in the area, n_2 is the number of animals captured in a sample from the region, and m_2 is the number of marked animals found in the sample of size n_2. This is then a valid procedure providing that the probability of an animal being included in the second sample of size n_2 is the same irrespective of whether or not it is radio-tagged. Furthermore, the standard error for the estimator from Equation (8.18) can be approximated using the usual equation

$$\widehat{SE}(\hat{N}^*) = \sqrt{\left[(n_1+1)(n_2+1)(n_1-m_2)(n_2-m_2)/\{(m_2+1)(m_2+2)\}\right]}. \tag{8.19}$$

If k independent surveys of the study area are made at different times to find animals, then each one of these surveys can be treated as the second sample for population estimation. The surveys will then provide k independent estimates $\hat{N}_1^*, \hat{N}_2^*, \ldots, \hat{N}_k^*$ of the population size, with corresponding estimated standard errors $\widehat{SE}(\hat{N}_1^*), \widehat{SE}(\hat{N}_2^*), \ldots, \widehat{SE}(\hat{N}_k^*)$.

The appropriate treatment of the series of population size estimates will depend on whether the population is open or closed. If the population is open, then the series of estimates can be used to monitor the size changes. This should give better results than the JS method because the number of marked animals available for capture is known instead of being estimated. On the other hand, if the population is closed, then all of the estimates should be made using the same fixed population size, and it is appropriate to think of producing some average of them to give the best combined estimate.

A number of methods have been proposed for producing a single combined estimate of the size of a closed population from a series of independent estimates. If all of the independent estimates are about equally precise, then the simple average

$$\hat{N}^* = (\hat{N}_1^* + \hat{N}_2^* + \ldots + \hat{N}_k^*)/k$$

is appropriate, with a standard error that can be estimated by the usual equation $\widehat{SE}(\hat{N}^*) = \sqrt{\{\Sigma_i \text{Var}(\hat{N}_i^*)/k\}}$. However, if this is not the case because the surveys were more intensive on some occasions than others, then a more complicated type of combined estimate should be used. White and Garrott (1990, Chapter 10) discussed six possible combined estimates and concluded that a joint maximum likelihood estimator is generally best, followed by the simple average.

8.6 Flexible Modeling Procedures

An approach to modeling mark-recapture data that has been advocated by Lebreton et al. (1992) represents a generalization of the work of Cormack (1964), Jolly (1965), and Seber (1965). First, a model is proposed to obtain estimates of capture and survival rates. The likelihood function for a set of data is then constructed by multiplying together the probabilities of observing the recapture patterns, given the time of first capture times. This function is then maximized with respect to the unknown parameters in the model. For example, consider a study of an animal population that lasts 7 years and suppose that the captures and recaptures of an animal are recorded as 0110100, where 1 indicates a capture and 0 indicates no capture. This animal is then recaptured in years 3 and 5 after its first capture in year 2. For such an animal, the probability of the recapture pattern is assumed to be

$$\phi_2 p_3 \phi_3 (1 - p_4) \phi_4 p_5 [(1 - \phi_5) + \phi_5 (1 - p_6)(1 - \phi_6) + \phi_5 (1 - p_6)\phi_6 (1 - p_7)],$$

where ϕ_i is the probability of surviving from year i to year $i + 1$, p_i is the probability of being captured in year i, and the three terms that are added within the square brackets are (1) the probability of dying in the year following the last sighting; (2) the probability of surviving to year 6, not being captured in that year, and dying in the following year; and (3) the probability of surviving to year 7 but not being captured in either year 6 or year 7. For a given model, probabilities can be obtained for all other recapture patterns in a similar manner, and the construction of the likelihood function is therefore in principle straightforward.

An important feature of this approach for modeling is the focus on the estimation of survival and capture probabilities instead of population sizes. One justification for this is that mark-recapture estimation works better for estimating these parameters than it does for the estimation of population sizes. Another justification is that it is survival probabilities that are important for population dynamics, and population size is just the outcome of survival and reproduction rates.

Of course, in some cases it really will be estimates of population size that are needed. These can, however, always be derived using the estimation equation $\hat{N}_i = n_i / \hat{p}_i$, where N_i is the estimated population size, n_i is the number of animals captured, and $\hat{p}_i$ is the probability of capture, all for the ith sample. This estimator is valid, providing that the probability of capture is the same for marked and unmarked animals, which is an assumption that is not required for the estimation of survival probabilities. The variance can then be estimated using an equation provided by Taylor et al. (2002) and discussed in Chapter 9 of Amstrup et al. (2005, p. 244).

8.6.1 Models for Capture and Survival Probabilities

In Chapter 7, in the section on the Huggins (1991) models, the advantages of modeling the probabilities of capture by means of a logistic function of the form

$$p_i = \exp(u_i)/\{1 + \exp(u_i)\}. \tag{8.20}$$

were discussed. Those same advantages apply here to both capture and survival probabilities of the form

$$\phi_i = \exp(v_i)/\{1 + \exp(v_i)\} \tag{8.21}$$

Incorporating these logistic functions of external covariates into the CJS likelihood was formally proposed by Lebreton et al. (1992) and makes it easy to model effects associated with time or individual differences because of age, weight, sex, and so on. For example, if x_i is a measure of the severity of the weather from year j to year $j + 1$ and y_i is a measure of the effort put into recapturing animals in year j, then it may be considered appropriate to model the probabilities of capture and survival by

$$\phi_j = \exp(\alpha_0 + \alpha_1 x_j)/\{1 + \exp(\alpha_0 + \alpha_1 x_j)\} \tag{8.22}$$

and

$$p_j = \exp(\beta_0 + \beta_1 y_j)/\{1 + \exp(\beta_0 + \beta_1 y_j)\}. \tag{8.23}$$

Once the α's and β's have been estimated (via maximum likelihood), the estimated survival and capture probabilities for any animal can then be calculated using the logistic equations.

One problem with using logistic regression functions with modeling survival probabilities is that there is no simple way to take into account changes in the time involved. For example, if there are 2 years between some samples and only 1 year between others, then this cannot be allowed for in a simple way by introducing into the model a parameter of time between samples. A better function to use in this respect would be the proportional hazards function $\phi_j = \exp\{-\exp(-u_j)t_j\}$ where t_j is the time between when samples j and $j+1$ are taken. So far, this approach to modeling survival does not seem to have been used. The approach usually taken is to model ϕ raised to the power t_j as a logistic function of covariates, then report the t_jth root of this parameter. This technique assumes survival of a period is a power function of the base survival (survival of 1 time unit) raised to the interval length (i.e., ϕ_j^t). This model may or may not be valid.

Some care is needed in setting up models for which parameters vary with time. For example, consider the logistic functions of Equations (8.22) and (8.23) for survival and capture probabilities in the CJS model with k samples. This model has $2k-2$ parameters $\phi_1, \phi_2, \ldots, \phi_{k-1}$ and $p_2, p_3, \ldots, p_k$, but it is only possible to estimate the product $\phi_{k-1}p_k$, rather than separate values for ϕ_{k-1} and p_k. One way to handle this within the logistic framework involves setting $\phi_{k-1} = \phi_{k-2}$ and using the parameterization

$$\phi_j = \begin{cases} \exp(\alpha_0 + \alpha_j)/\{1+\exp(\alpha_0 + \alpha_j), & 1 \le j \le k-3 \\ \exp(\alpha_0)/\{1+\exp(\alpha_0), & j = k-2 \text{ and } k-1 \end{cases}$$

and

$$p_j = \begin{cases} \exp(\beta_0)/\{1+\exp(\beta_0), & j = 2 \\ \exp(\beta_0 + \beta_j)/\{1+\exp(\beta_0 + \beta_j), & 3 \le j \le k \end{cases}$$

The $2k-3$ parameters in this case are $\alpha_0, \alpha_1, \ldots, \alpha_{k-3}$ and $\beta_0, \beta_3, \beta_4, \ldots, \beta_k$, which can all be estimated. Other alternative parameterizations would be equally good in the sense that they would give the same estimates of $\phi_1, \phi_2, \ldots, \phi_{k-2}$ and $p_2, p_3, \ldots, p_{k-1}$.

This type of complication occurs with other models as well. It is therefore important to know which parameters can be estimated from the available data and adopt an appropriate set of parameters for the logistic functions. Unfortunately, the variable metric algorithm for maximizing the likelihood function that is currently favored (see the following discussion) is capable of determining this maximum even when the model being fitted contains more parameters than it is really possible to estimate separately. An inappropriate parameterization may therefore not be obvious just from the results of the fitting process.

8.6.2 Possible Candidate Models

The type of approach that is used for analyzing data involves defining a set of candidate models from which one will be chosen. For example, if the sex is recorded for each individual that is marked, then the following models might be entertained for capture probabilities: (1) sex*time, for which the capture probability varies with the sample time and differs for males and females; (2) sex + time, for which the capture probability varies with the sample time and the constant term in the logistic model also varies for males and females; (3) sex, for which the capture probability is constant over time but is different for males and females; (4) time, for which the capture probability varies with time but is the same for males and females; (5) trend, for which there is a trend in capture probabilities; and (6) constant, for which capture probabilities are constant.

These models can be expressed using the logistic function of Equation (8.20) through judicious choice of covariates. Therefore, for example, the model sex*time is set up by making the argument u_i of this function equal to a sum of suitable indicator variables for the effects of time, sex, and the interaction between these factors, as discussed by Lebreton et al. (1992, p. 77). On the other hand, the model sex + time only includes variables for the main effects of sex and time in the sum u_i.

The models time and trend both allow the survival probability to vary with time. The difference is that with the time model the value u_i of Equation (8.20) includes a component that is different for every sample time, whereas with the trend model the sample time is included as a quantitative variable with a coefficient to be estimated. This means that with the trend model there is only one parameter to be estimated to account for time changes, but the time model requires several parameters. Furthermore, the trend model is a special case of the time model.

The six models for survival probabilities would probably be the same as those used for capture probabilities. By considering all different combinations of the six capture probability models and the six survival probability models, there are then $6 \times 6 = 36$ possible models to be considered for the data. It must be stressed that these 36 models may or may not be suitable for a particular mark-recapture study.

8.6.3 Maximum Likelihood Estimation

Most models for capture–recapture data require the numerical maximization of the likelihood function. Many different algorithms can be used for this purpose, but variable metric methods have the advantage of not requiring the calculation of second derivatives of the likelihood function. Furthermore, if the function maximized is the log-likelihood rather than the likelihood itself, then these algorithms output an approximation to the Hessian matrix, which in our application is the covariance matrix for the estimated parameters. SURGE (Pradel and Lebreton, 1991) includes an algorithm of this type, and for the program MRA-LGE (Supplementary Material for this chapter available at https://sites.google.com/a/west-inc.com/introduction-to-ecological-sampling-supplementary-materials/) the implementation of this algorithm in the FORTRAN subroutine DFMIN provided by Press et al. (1992) has been used. The 'mra' package (http://cran.r-project.org/web/packages/mra/index.html) associated the R computer language, and MARK (http://www.phidot.org/software/mark/ downloads) also implement this type of algorithm.

8.6.4 Likelihood Ratio Tests to Compare Two Models

One of the useful tools that is available when maximum likelihood estimation is used is a test for whether one model is a significantly better fit than an alternative simpler model. To use this test, it is necessary that the first model has more parameters than the second model and includes the second model as a special case. That is, by constraining one or more of the parameters of the first model in some way, the second model is obtained.

Let L_1 be the value of the maximum of the likelihood function for the first model, which has p_1 estimated parameters, and L_2 be the maximum of the likelihood function for model 2, which has p_2 estimated parameters. Then, $p_1 > p_2$, and it can be anticipated that $L_1 > L_2$ because of the extra parameters in model 1. A standard result, then, is that if the simpler model 2 is in fact correct, the statistic

$$D = 2\{\log_e(L_1) - \log_e(L_2)\} \tag{8.24}$$

will approximately have a chi-squared distribution with $p_1 - p_2$ degrees of freedom (df).

8.6.5 Model Selection Using the Akaike's Information Criterion

In recent years the Akaike's (1973) information criterion (AIC) has become popular for selecting models for mark-recapture data (Anderson et al., 1994; Burnham and Anderson, 1992, 2002; Burnham et al., 1994; Huggins, 1991; Lebreton et al. 1992; Lebreton and North, 1993). Burnham et al. (1995) have shown, through a simulation study, that this criterion gives a good balance between bias caused by using models with too few parameters and high variance caused by using models with too many parameters.

The criterion is calculated as

$$AIC = -2 \log_e(L_{max}) + 2K, \tag{8.25}$$

where L_{max} is the likelihood for the data evaluated at the point where it is maximized, and K is the number of estimated parameters for a model. The AIC value is found for a range of plausible models, and the model selected is the one for which AIC is a minimum. Models with many parameters are penalized by the term $2K$. To be selected, they must have a much higher likelihood than models with a small number of parameters.

8.6.6 Overdispersion

Overdispersion is a well-known phenomenon with count data. It occurs when the variation in the data is more than can be expected on the basis of the model being considered. With mark-recapture data, overdispersion can arise because the recapture patterns are not independent for different animals or because parameters that are assumed to be constant are really varying.

The simplest way to allow for overdispersion involves assuming that all variances are increased by the same variance inflation factor (VIF) c, but the assumed model is otherwise correct. Then, c can be estimated from the data, and simple adjustments can be made to various types of analysis. In the mark-recapture context, this can be done by taking the estimate

$$\hat{c} = X^2/df \qquad (8.26)$$

where X^2 is the overall test statistic obtained from the goodness-of-fit TEST 2 and TEST 3 (described in the next section), with df degrees of freedom.

An explanation for why the use of a VIF may be effective is based on the idea of what would happen if a data set was artificially increased in size by cloning. If the mark-recapture pattern of each animal is entered into the data set twice instead of once, then all of the variances of parameters will be halved and $\hat{c}$ will be doubled in comparison with what is obtained for the original data set. More generally, if each animal is entered into the data set R times, then all of the variances will be divided by R and $\hat{c}$ will be multiplied by R in comparison with what is obtained from the original data set. The VIF can therefore be interpreted as the amount by which the size of a data set is larger than it would be for a set for which all animals provide independent results. The use of Equation (8.26) to estimate the VIF is justified by the fact that the expected value of $\hat{c}$ is approximately one if the assumptions of the fitted model are correct and all animals provide independent data. Therefore, if the assumptions of the model are correct but the animals do not provide independent data, then the expected value of $\hat{c}$ is c.

In reality, the reasons for overdispersion will usually be much more complicated than just the duplication of the results for individual animals. Nevertheless, it can be hoped that use of $\hat{c}$ will be effective in allowing for heterogeneity in data in cases for which this is needed.

Anderson et al. (1994) showed that use of $\hat{c}$ seems to work well, with a good balance between over- and underfitting of models when these are selected on the basis of the quasi-AIC (QAIC)

$$\text{QAIC} = -2 \log_e(L_{max})/\hat{c} + 2K. \tag{8.27}$$

Apart from its use with model selection, the VIF can also be used to adjust the variances of parameter estimates. This can be done by multiplying the variances that are obtained from the usual model-fitting process by $\hat{c}$.

If an adjustment using a VIF is used on a routine basis, then it should be realized that the correct value will generally be one or more. Consequently, if Equation (8.26) gives $\hat{c} < 1$, then this should be changed to $\hat{c} = 1$; that is, it should be assumed that there is no overdispersion.

8.7 Tests of Goodness of Fit

Burnham et al. (1987) discussed three tests for the goodness of fit of models for mark-recapture data. TEST 1 is applicable when there are two or more groups of animals receiving different treatments. This is not considered further here because the test cannot be applied with most sets of data. TEST 2 tests the null hypothesis that survival and capture probabilities are the same for all animals for a particular sample or period between samples, that is, that the JS model is correct. TEST 3 also tests whether the assumptions of the JS model are correct, but against the alternative hypothesis that survival and capture probabilities vary with the time of release. See the work of Burnham et al. (1987) for more information about these tests.

8.8 An Example of Mark-Recapture Modeling

Examples of the use of the methods that have been described are now considered. The calculations needed for these examples were carried out using three computer programs, MRA-LGE, MRA36, and the mra package for R (see the Supplementary Material for this chapter), although other programs such as MARK or routines in the RMark packages for R could have been used instead. Basically, MRA-LGE fits any mark-recapture model to data where the probabilities of survival and capture can be expressed through logistic functions of the form of Equations (8.20) and (8.21), and MRA36 fits the 36 models described in Section 8.6 to data with two groups of animals, such as males and females, or fits the relevant subset of nine models if there is only one group of animals. R code necessary to fit the 36 models described in Section 8.6 is available on the web site.

One of the examples considered by Lebreton et al. (1992) concerned the European dipper (*Cinclus cinclus*) in eastern France. In this case, there are mark-recapture records for 294 birds, consisting of 141 males and 153 females. There were seven samples taken at yearly intervals, from 1981 to 1987, and there is the additional information that the survival probabilities for years 1982–1983 and 1983–1984 may have been affected by a major flood.

8.8.1 Constant Probability of Capture

Consider the fitting of the model which posits that the probability of a marked bird being recaptured was the same for each of the years 2 to 7 when this was possible, and the probability of surviving from one year to the next varied with time but was the same for males and females. This is the model (p constant, ϕ time). It must be stressed that this model is only used here to illustrate the model-fitting process. It would not be an obvious choice for the first model to fit to these data at the start of a serious analysis.

For this model, the estimated constant probability of capturing a marked bird in any year is $p = \exp(2.220)/\{1 + \exp(2.220)\} = 0.902$, with an estimated standard error of 0.029, while the estimated survival rates for years 1 to 6, with standard errors in parentheses, are

$$\phi_1 = \exp(0.336 + 0.178)/\{1 + \exp(0.336 + 0.178)\} = 0.626\ (0.113)$$

$$\phi_2 = \exp(0.336 - 0.520)/\{1 + \exp(0.336 - 0.520)\} = 0.454\ (0.067)$$

$$\phi_3 = \exp(0.336 - 0.423)/\{1 + \exp(0.336 - 0.423)\} = 0.478\ (0.059)$$

$$\phi_4 = \exp(0.336 + 0.172)/\{1 + \exp(0.336 + 0.172)\} = 0.624\ (0.058)$$

$$\phi_5 = \exp(0.336 + 0.102)/\{1 + \exp(0.336 + 0.102)\} = 0.608\ (0.055)$$

and

$$\phi_6 = \exp(0.336)/\{1 + \exp(0.336)\} = 0.583\ (0.058).$$

These estimates relate to the survival from 1981 to 1982, 1982 to 1983, and so on up to the survival from 1986 to 1987.

There are seven estimated parameters in this model. The maximized log-likelihood is −329.87, and the AIC of Equation (8.25) is 673.73. TEST 2 and TEST 3 give an overall statistic of 11.76 with 10 df, which is not at all significantly large. There is therefore no evidence of lack of fit of the basic JS model for these data.

8.8.2 Allowing for the Effect of a Flood

Suppose now that there is some interest in comparing the model just fitted with a simpler one that says that the probability of recapture was the same for all birds in all years but the probability of survival was one of two values. For the periods 1982–1983 and 1983–1984, there was one survival probability because of the effects of the flood, and a different survival probability applied for the other years.

This model can be fitted using MRA-LGE when a *Y* variable is included in the input data, which is 1 for the years affected by the flood and otherwise 0. Fitting the three-parameter model with a constant probability of capture and two survival rates leads to a maximized log-likelihood of -330.05. The log-likelihood test to compare the fit of this model with the one previously fitted then gives $D = 2\{(-329.87) - (-330.05)\} = 0.36$, from Equation (8.24), with $7 - 3 = 4$ df. On this basis, there is no evidence that the extra complexity of the first model is needed.

8.8.3 Model Selection Using the QAIC

The computer program MRA36 can be used to fit the 36 models that were defined in Section 8.6 to the data on the European dipper. It was found that $\hat{c}$ equaled 1.18 and the best model according to the QAIC is (*p* constant, ϕ constant), that is, the probability of capture and the probability of survival were the same for all birds at all times. Because the model allowing for a flood effect is not one of the ones considered by MRA36, there is now some interest in comparing the (constant, constant) model with the flood effect model. It is then found that the QAIC suggests that the model with a flood effect allowed is the most appropriate of the two considered.

8.9 Recent Advances with Open-Population Models

The book *Handbook of Capture–Recapture Analysis* (Amstrup et al., 2005) discusses many recent advances concerning the analysis of capture–recapture data for open and closed populations, tag-recovery models, the joint analysis of tag-recovery and live-resighting data, and multistate models (in which animals move between several states while capture–recapture sampling takes place). It also includes the discussion of the analysis of eight data sets.

8.10 General Computer Programs for Capture–Recapture Analyses

The *Handbook of Capture–Recapture Analysis* also contains information about 11 programs that are available for the analysis of capture–recapture data. Of these, MARK is the best-known program devoted to capture–recapture analyses, and it includes the capabilities of most of the other programs. It is available free on the Internet (http://www.phidot.org/software/mark/downloads) and includes a comprehensive guide. Some of the examples in the handbook also illustrate the use of this program. A healthy discussion forum for the MARK program, and capture–recapture analyses in general, is available at phidot.org (http://www.phidot.org/forum/index.php).

In R, two options for fitting capture–recapture models exist. The first is routines in the RMark package (http://cran.r-project.org/web/packages/RMark/). These routines allow the use of the R language to manipulate and plot data, while MARK is called in the background to estimate models. In this way, RMark is an interface to the estimation routines in MARK. The other R option is to use the routines in the mra package (http://cran.r-project.org/web/packages/mra/). Like RMark, these routines allow the use of R to manipulate and plot data, but unlike RMark, these routines estimate models using FORTRAN routines behind the scenes.

9

Occupancy Models

Darryl MacKenzie

9.1 Introduction

Occupancy models are a set of techniques for investigating the presence-absence of a species while accounting for the fact that the species may be present but goes undetected by the survey methods used at some locations. Such methods can be useful in a wide range of ecological applications, such as species distribution modeling, metapopulation studies, habitat modeling, and resource selection functions, and can also be useful for monitoring programs (MacKenzie et al., 2006). They are useful not only for assessing patterns in species occurrence, such as identifying important habitat relationships to the current distribution, but also for understanding and predicting changes in species distributions through time. Occupancy models have also been extended beyond presence-absence data (or, more generally, two categories for the status of the species at a unit) to multiple categories (e.g., absence and presence with or without breeding; or none/some/many individuals) (Royle and Link, 2005; Nichols et al., 2007; MacKenzie et al., 2009).

Study design is a key aspect for the correct application of occupancy models. Without reliable data, inferences from occupancy models, as from any modeling exercise, might be unreliable. Through a careful design process, conclusions resulting from the modeling are likely to be more accurate, and realistic expectations about the success of the study or monitoring program can be set.

This chapter reviews some of the main features of occupancy models from both analysis and design perspectives. The chapter begins with a general overview of the type of data and basic sampling requirements, followed by introductions to models that can be used to address questions concerned with species occurrence during a single period of time (single-season models) and changes in occurrence (multiseason models). Study design issues are then discussed, and the chapter ends with a general discussion.

9.2 General Overview

A situation is envisioned for which the occurrence of a target species is of interest within some predefined region. This region consists of a number of units at which the species may be either present or absent (or otherwise categorized). What constitutes a unit will depend on the target species and objectives of the study or monitoring program. The units may be defined in terms of natural features such as ponds or patches of habitat or arbitrarily defined as grid cells, stream segments, transects, and so on. The collection of all units within the region is considered the statistical population of interest. These issues are discussed in further detail later in this chapter.

The status of the species at each unit can be one of multiple categories at a particular point in time. Examples of such categorizations include presence/absence; used/unused; presence with breeding/presence without breeding/absence. The unit's status may change over time through certain dynamic processes. It is presumed that there is a period of time when the status of the species can be regarded as stable or static. Here, that time period is referred to as a *season*, which may or may not correspond to a biologically relevant season such as a climatic or breeding season.

Surveys are conducted to determine the current status of the species at the units. It is not necessary that all units within the region of interest be surveyed as a sampling scheme can be used to select a subset of units to survey, with the results generalized to the entire population of units. The total number of units within the region of interest is denoted here as S, and the size of the sample from this population is s.

A practical consideration is that often the field methods used to survey for the species in the units will be imperfect. That is, the true status of the species at the unit will not always be observed, and there is the potential for misclassifying the unit. It is assumed that any potential misclassification will be one way; that is, there are some types of observations for which the species status can be definitively determined, while other observations are associated with potential ambiguity. For example, detection of the species confirms the species is present, but nondetection does not confirm the species is absent.

Because of imperfect detection, multiple surveys for the species should be conducted within each season. The exact nature of the repeat surveys can vary with different applications, but options include discrete visits to each sampling unit or multiple surveys conducted within a single visit. Further suggestions are made in the section on study design.

9.3 Single-Season Models

Single-season models can be used when interest is in how the species is distributed on the landscape at a single time point. Patterns in the current distribution could be described and related to covariates or predictor variables that are available for each sampling unit. Initially, the single-season model is developed in terms of a two-category occupancy model (e.g., presence-absence), then an extension to multiple categories is discussed.

9.3.1 Two-Category Model

In a two-category application, the status of the species at each sampling unit may be one of two possible categories. Typical examples are the presence or absence of the species from each unit or whether a unit is used or unused by the species. Other definitions are also possible, for example, high abundance or not high abundance. Field observations from the surveys correspond to each of these categories (e.g., detection-nondetection), but while one type of observation would confirm the status of the species at the unit (e.g., the species would have to be present to be detected), there is ambiguity associated with the other observation (e.g., a nondetection could result both when the species was present or when it was absent).

The repeat surveys are necessary to reliably disentangle the species occurrence from the observation process. An example *detection history* from a single unit would be $h_i = 101$, with the 1 indicating the species was detected in that survey and a 0 indicating nondetection. Presuming the two categories of interest are presence-absence, a verbal description of the detection history would be that "the species was present at the unit, detected in the first survey, not detected in the second, and then detected in the third."

Let ψ be the probability of the species being present at a unit and p_j be the probability of detecting the species in the jth survey of the unit given the species is present. From the verbal description, an expression for the probability of observing the detection can be developed by substituting the respective phrases for the associated probability. Here, the resulting expression is referred to as a *probability statement*. The probability statement for the previous detection history at unit i (h_i) would therefore be

$$\Pr(h_i = 101) = \psi p_1 (1 - p_2) p_3.$$

If the species was never detected in any of the three surveys (i.e., 000), then there is ambiguity regarding whether the species was truly present or absent from the unit. Again, a verbal description of that detection history would be that "the species was present but went undetected in all three of the surveys, or the species was absent." While it is not possible to resolve this ambiguity based on the given data, it can be accounted for within the probability statement to remove any bias caused by the imperfect detection. That is, the probability statement for this detection would be

$$\Pr(h_i = 000) = \psi(1 - p_1)(1 - p_2)(1 - p_3) + (1 - \psi).$$

The addition of the two terms in the probability statement is how the ambiguity in the observation is accounted for; each term represents the possible explanation for how the species could never be detected at a unit, either a false or a genuine absence, respectively.

Probability statements can be developed for all s sampling units that are surveyed, and in combination can either be used within a maximum likelihood or Bayesian framework to obtain results in the usual way. Note also that equal sampling effort at each unit is not required so that the number of surveys per unit need not be consistent (see MacKenzie et al., 2002).

EXAMPLE 9.1 Blue-Ridge Two-Lined Salamander

MacKenzie et al. (2006) presented results from data collected on the blue-ridge two-lined salamander (*Eurycea wildrae*) in the Great Smoky National Park, Tennessee. Thirty-nine 50-m transects were surveyed on five occasions between April and mid-June 2001. Each transect consisted of one natural cover transect where objects such as rocks and logs were lifted to check for salamanders and a parallel artificial cover object transect where cover boards (pine boards) were placed every 10 m and searched to reduce the impact on the environment. MacKenzie et al. (2006) pooled the data from the two types of transect to obtain a single detection or nondetection for each survey of a transect. Further details of the original study can be found in the work of Bailey et al. (2004).

Blue-ridge two-lined salamanders were detected at least once at 18 of the 39 transects. This would correspond to an uncorrected estimate of the probability of salamanders being present on a transect of 0.46 with a standard error (SE) of 0.11. Using the methods described to account for the possibility of false absences, assuming the detection probability was the same for all five surveys, the estimated probability of salamanders being present on a transect is 0.59 with an SE of 0.12. The nearly 30% increase of the uncorrected estimate is a result of detection probability being estimated to be low at 0.26 (SE = 0.05) per survey; hence, the probability of not detecting the salamanders on a transect where they were present after five surveys would be 0.22 [i.e., $(1 - 0.26)^5$]. That is, there was a 22% chance of not detecting salamanders along a transect after five surveys even when they were present.

9.3.2 Multiple-Category Model

The two-category model of MacKenzie et al. (2002) was extended by Royle and Link (2005) and Nichols et al. (2007) for situations when two categories for the status of the species across different points of the landscape are insufficient. Royle and Link (2005) considered a situation for which four categories of relative abundance were of interest based on call-index data for breeding amphibians. Nichols et al. (2007) considered a situation for which three categories were of interest for assessing the amount of successful reproduction of California spotted owls (*Stryx occidentalis occidentalis*) at a study site in the U.S. Sierra Nevada mountains, where the categories of interest were absent, present without successful reproduction, and present with successful reproduction. In each case, there was potential for one-way misclassification. For example, registering calls from a few individuals precludes the possibility of the species being absent, but a greater number of individuals may have been present at the sample unit with some going undetected. Although Royle and Link (2005) and Nichols et al. (2007) used different parameterizations, the underlying modeling framework was the same, as noted by MacKenzie et al. (2009). Details of the different parameterizations are not given here, and readers are directed to the articles cited. A rather general parameterization is used in the following material, but it is noted that other parameterizations can be used and are likely to be more reasonable for specific applications.

In the multiple-category situation, the outcome of each survey can be recorded as a 0, 1, 2, and so on, rather than just a 0 or 1 as for the two-category situation. A similar approach to the two-category model is used by which any ambiguity in the detection history is resolved by adding together the probabilities to the individual outcomes. Technically, this procedure is known as integrating across the possible occupancy states. For example, if the parameters were defined as

$\phi^{[m]}$ is the probability of a unit being of category m, and

$p_j^{[l,m]}$ is the probability of observing evidence of category l in the jth survey, given the unit is truly of category m,

the probability statement for the detection history 021 would be

$$\Pr\left(h_i = 021\right) = \phi^{[2]} p_1^{[0,2]} p_2^{[2,2]} p_3^{[1,2]}.$$

For example, in the context of the three categories being the presence of the species with and without breeding, the verbal description of this detection history is the following: "The species is present at the unit and breeding is occurring there; the species was not detected at all in the first survey; it was detected in the second survey, and the evidence of reproduction was also observed; then in the third survey the species was detected but there was no evidence of reproduction." In this case, there is no ambiguity associated with the detection history as the evidence of breeding (i.e., a 2) was observed at least once during the surveys, confirming that breeding was occurring.

A second example for which there is ambiguity is the outcome 110. Then, the detection of the species at least once precludes the possibility of the species being absent from the unit, but not detecting evidence of reproduction does not preclude the possibility of reproduction actually occurring there. The probability statement for this would be

$$\Pr\left(h_i = 110\right) = \phi^{[2]} p_1^{[1,2]} p_2^{[1,2]} p_3^{[0,2]} + \phi^{[1]} p_1^{[1,1]} p_2^{[1,1]} p_3^{[0,1]}.$$

Here, the two options relate to whether reproduction is occurring at the unit or not. That is, the species may have been present with reproduction, the species was detected but no evidence of reproduction was observed in the first and second surveys, and the species was not detected in the third survey, *or*, the species was present but there was no reproduction at the unit and the species was detected in the first and second survey but not in the third survey. Note that in the second option it is presumed that there is no reproduction occurring at the unit; therefore, there is no chance of observing it. Hence, the second superscript on the detection probabilities is a 1 (species is present without reproduction).

As before, once the probability statement for all sampled units has been determined, the statements can be combined and used to provide maximum likelihood estimates or used within a Bayesian framework to obtain posterior distributions for the parameters.

9.4 Multiseason Models

When changes in the distribution and occurrence of the species are of interest, then multiseason occupancy models can be used. These can be developed such that the current occupancy state of a sampling unit is independent of its state in the previous time point or by allowing a dependency to exist between the occupancy status of the unit in the two time periods. In this section, the independent modeling approach is briefly outlined before moving on to the dependent case, which is likely to be more realistic in many situations. The modeling is described in terms of a multicategory situation, of which the two-category situation (e.g., presence-absence) is just a special case.

9.4.1 Independent Changes

Just because the following modeling approach assumes that the current status of a unit is independent of its state in the previous time period does not mean the model cannot be used when there really is dependence. In terms of modeling the overall patterns of occurrence each time period, using a model that assumes independence still provides appropriate results. What it fails to do, however, is to provide insight into the underlying processes of change as it essentially assumes that any changes in occupancy are random (MacKenzie et al., 2006).

By assuming independence, a multiseason model can be constructed by fitting a series of single-season models to each season of data. Changes in the overall patterns of occurrence (such as trends over time) can be incorporated by creating a functional link between the parameters for each season. For example, when interest is in the presence or absence of the species, the probability of occupancy in each year t could be modeled as

$$\text{logit}\left(\psi_t\right) = \beta_0 + \beta_1 \cdot t,$$

where β_1 is the trend in occupancy measured on the logit scale (as discussed further in the section on incorporating covariates).

Although it is advisable to survey the same units each time period, when this has not happened, then the data collected in each season might be relatively independent (e.g., if the units to be surveyed are randomly selected each season). It is therefore unlikely that there will be data from consecutive time periods on the same unit. Hence, in this situation, this modeling is likely to be more successful than the next approach, which specifically models the underlying changes at the sampling unit scale.

9.4.2 Dependent Changes

MacKenzie et al. (2003) extended the two-category, single-season model of MacKenzie et al. (2002) to the situation in which data are collected at systematic points in time at the same sampling units, hence providing data of a longitudinal nature. As for the single-season model, at each time point repeat surveys are conducted to provide information about detection probability. The model accounts for changes in the occupancy status of units between seasons through the processes of local colonization and extinction. Although different terms could be used for them, basically they enable the probability of the species being present at a unit to be different depending on whether the species was present at the unit in the previous season. Barbraud et al. (2003) described a similar approach developed from mark-recapture models. For multiple categories, MacKenzie et al. (2009) extended the single-season models of Royle and Link (2005) and Nichols et al. (2007) to the multiple-season situation, again assuming that the data are of a longitudinal nature.

Probability statements can again be developed by adding together various options to account for any ambiguity in the true state of a unit caused by the imperfection of the observations. This is most efficiently achieved by using matrix multiplication with appropriately defined vectors and matrices, and the key to the modeling is a transition probability matrix. For full details, see the work of MacKenzie et al. (2003, 2006, 2009).

A transition probability matrix simply defines the probability of a unit being in a particular category in season $t + 1$ conditional on the category of the unit in season t. This conditional approach creates the dependence structure of the model; technically, the modeling is assuming a first-order Markov process. If the probability of a unit changing from state m in season t to state n in $t + 1$ is $\phi_t^{[m,n]}$, then the transition probability matrix for a situation with three possible categories, for example, would look like the following:

$$
\phi_t = \begin{bmatrix}
\phi_t^{[0,0]} & \phi_t^{[0,1]} & \phi_t^{[0,2]} \\
\phi_t^{[1,0]} & \phi_t^{[1,1]} & \phi_t^{[1,2]} \\
\phi_t^{[2,0]} & \phi_t^{[2,1]} & \phi_t^{[2,2]}
\end{bmatrix},
$$

where the rows relate to the category of the unit at time t and columns denote the category at $t + 1$. Note that each row has to sum to 1.0; therefore, two of the three probabilities in each row would be estimated and the third determined by subtraction. In this form, the parameterization is multinomial, meaning that the two estimated probabilities are independent and unconstrained (except for the fact their sum must be less than 1.0.

Sometimes, the multinomial parameterization can have numerical difficulties, particularly when covariates are to be included, and the interpretation of covariates effects can also be unclear. When it makes sense, an alternative parameterization is to use a conditional binomial approach. For example, if $\psi_{t+1}^{[m]}$ is defined as the probability of the species being present at the unit at time $t + 1$ conditional on the unit existing in category m at time t, and $R_{t+1}^{[m]}$ is defined as the probability of reproduction occurring at a unit at time $t + 1$ given the species was present at the unit at $t + 1$, and the unit was in category m at time t, then the transition probability matrix would look like this:

$$
\phi_t = \begin{bmatrix}
1 - \psi_{t+1}^{[0]} & \psi_{t+1}^{[0]}\left(1 - R_{t+1}^{[0]}\right) & \psi_{t+1}^{[0]} R_{t+1}^{[0]} \\
1 - \psi_{t+1}^{[1]} & \psi_{t+1}^{[1]}\left(1 - R_{t+1}^{[1]}\right) & \psi_{t+1}^{[1]} R_{t+1}^{[1]} \\
1 - \psi_{t+1}^{[2]} & \psi_{t+1}^{[2]}\left(1 - R_{t+1}^{[1]}\right) & \psi_{t+1}^{[2]} R_{t+1}^{[2]}
\end{bmatrix}
$$

Essentially, this parameterization is analogous to flipping a series of coins to determine which category a unit is in rather than rolling a die.

Note also that biologically interesting questions can be considered with these modeling approaches. For example, does the probability of reproduction occurring at a unit depend on whether reproduction occurred at the unit previously? If so, that would imply $R_{t+1}^{[1]} \neq R_{t+1}^{[2]}$; if not, then $R_{t+1}^{[1]} = R_{t+1}^{[2]}$. Such questions could be relevant in identifying potential costs of reproduction or source-sink dynamics.

9.5 Including Covariates

Any of the probabilities in both the single- and multiseason models could be modeled as a function of covariates or predictor variables. Indeed, often it will be these relationships that are of primary interest to explore which landscape features are important for the distribution of the species. However, while the covariate may potentially take any positive or negative value, a probability must be on the scale of 0–1. Therefore, a transformation, or *link function*, must be used to ensure that the probabilities and covariates are on a comparable scale. There are a range of options available for doing so, including the probit link, the log-log link, and the complementary log-log link, although the one discussed here is the logit link, which is the basis of logistic regression and is a commonly used technique for analyzing binary data. The logit link function is defined as

$$\text{logit}\left(\theta_i\right) = \ln\left(\frac{\theta_i}{1-\theta_i}\right) = \beta_0 + \beta_1 \cdot x_{1,i} + \beta_2 \cdot x_{2,i} + \dots + \beta_r \cdot x_{r,i}$$

where θ_i is the probability of interest for unit i, $x_{1,i}$ to $x_{r,i}$ are the values of the covariates of interest measured for unit i, and β_0 to β_r are the regression coefficients to be estimated. Note that the ratio $\theta_i/(1-\theta_i)$ is the odds of the event occurring; hence, the logit link is also referred to as the log-odds link. For the probabilities used in the multinomial parameterizations, the multinomial-logit link should be used.

Note that the logit link is a form of a generalized linear model (GLM); hence, users need to specify the functional form of the relationship between the probability and the covariate (e.g., linear, quadratic, etc.). This is not to say that alternative approaches such as generalized additive models (GAMs) cannot be used with occupancy models for more flexible curve-fitting relationships with covariate.

There are two broad types of covariates that could be considered with occupancy models. First, unit-specific covariates such as vegetation type, distance from water, elevation, and so on can be used as covariates for all the probabilities discussed previously, both occupancy related and those associated with the detection process. These covariates are essentially characteristics of each unit and are assumed to be constant for a particular season, but they may change between seasons. Second, covariates with survey-specific values (e.g., the time of survey, wind conditions, air temperature, etc.)

can be used for the detection probabilities in addition to the unit-specific covariates.

Finally, it should be noted that both continuous and categorical covariates may be used. Continuous covariates could potentially take on any positive or negative value; categorical covariates can only have a number of specific values and will often need to be converted to a series of indicator or dummy variables. However, if the categories have been ranked in order and assigned meaningful numeric values, then the covariates can be considered as ordinal categorical covariates and are essentially treated as continuous covariates.

EXAMPLE 9.2 Spread of the House Finch across Eastern North America

To illustrate the usefulness of multiple-season models and the incorporation of covariates, a data set from the North American Breeding Bird Survey (BBS) for the house finch (*Carpodacus mexicanus*) is used. This data set was previously discussed by MacKenzie et al. (2006), and greater detail of the data set can be found in Section 7.3 of that book.

The house finch is native to western North America but not midcontinental and eastern North America. A small population was released in Long Island, New York, in 1942 and began expanding their range westward. The BBS conducts surveys along secondary roads during the peak of the breeding season, with observers traveling an approximately 39.2-km route of 50 equally spaced stops. At each stop, a 3-minute point count is conducted, and observers note which species they detected within a 400-m radius. Here, the BBS route is considered as the sampling unit and the 50 replicate stops as the repeat surveys to detect house finches if they are present along the route.

BBS data from 1976 to 2001, at 5-yearly intervals, were analyzed with a two-category (i.e., presence-absence) multiseason model with dependent changes to model the expansion of house finch range across eastern North America. The parameters in the model are the probabilities of occupancy in 1976, local colonization between periods t and $t + 1$, local extinction between periods t and $t + 1$, and the detection of house finches in a survey if they were present at a route in period t (MacKenzie et al., 2003, 2006). Data were extracted for 694 BBS routes that were surveyed in at least one of the 6 years of interest. Distance bands of 100 km from Long Island were defined (e.g., 0–99 km, 100–199 km, etc.) and considered as a potential covariate for all parameters in the model. MacKenzie et al. (2006) also defined a covariate to enable the probability of detecting house finches at a stop along a route to be different if house finches had been detected at greater than 10 stops along that route in a previous year. The logic of such a covariate was to attempt to account for any effect of local abundance on the detection of house finches (at the species level).

MacKenzie et al. (2006) considered a number of different models for the data, but for simplicity only the results from a single model are considered here. The initial probability of house finches being present at route i in 1976 $\psi_{76,i}$ was modeled as

$$\text{logit}(\psi_{76,i}) = a_1 + a_2 \cdot D_i,$$

where D_i is the distance band for route i scaled such that $D_i = 1$ for the 1000- to 1099-km distance band (i.e., essentially the distance in 1000s of kilometers from the Long Island release point). The probability of house finches colonizing route i between periods t and $t + 1$ ($\gamma_{t,i}$) was modeled with the effect of distance being different in each period, that is, an interaction between period and distance effects so that

$$\text{logit}(\gamma_{t,i}) = b_{t,1} + b_{t,2} \cdot D_i.$$

The probability of house finches going locally extinct from route i between periods t and $t + 1$ ($\varepsilon_{t,i}$) was modeled as a function of distance only, with no period effects, that is, with no annual variation in the extinction probability of

$$\text{logit}(\varepsilon_{t,i}) = c_1 + c_2 \cdot D_i.$$

The effect of distance on detection probability was also allowed to vary for each time period (as for colonization), with an additional effect for the local abundance that was consistent in all time periods. That is, the probability of detecting house finches at a stop along route i in time period t (given house finches present at route i) was modeled as

$$\text{logit}(p_{t,i}) = d_{t,1} + d_{t,2} \cdot D_i + d_3 \cdot LA_{t,i}$$

where $LA_{t,i} = 1$ if house finches were detected at more than 10 stops in a period previous to t and zero otherwise.

The resulting parameter estimates from the analysis of the data are given in Table 9.1 and suggest that the probability of house finch presence on a route in 1976 declined with the distance from Long Island ($\hat{a}_2 < 0$), and that the colonization probability also declined with distance, although the magnitude of the effects also (generally) declined through time (i.e., $\hat{b}_{t,2}$ were negative but became closer to zero as the time period increase). The extinction probability increased with distance ($\hat{c}_2 > 0$), and the effect of distance on detection exhibited a similar pattern to colonization probability over time. House finches were also more detectable at a stop if they had been detected at more than 10 stops along a route previously ($\hat{d}_3 > 0$). Plots of the estimated occupancy-related biological probabilities are given in Figures 9.1 to 9.3. Note that the estimated probabilities can be used to calculate other quantities, such as the probability of presence in each time period (Figure 9.4) or rate of change in occupancy (Figure 9.5) as a function of distance as alternative ways to describe the expansion of the house finch across eastern North America (MacKenzie et al., 2006).

Using the distance from Long Island (the initial source of the eastern population) as a covariate on all parameters and allowing the nature of that relationship to change through time was an attempt by MacKenzie

TABLE 9.1

Parameter Estimates and Standard Errors (SE) for the Model to Describe the Expansion of the House Finch across Eastern North America during 1976–2001 at Five-Yearly Intervals

	Parameter	Estimate	SE		Parameter	Estimate	SE
Occupancy	$\hat{a}_1$	−0.83	0.41	Extinction	$\hat{c}_1$	−3.39	0.26
	$\hat{a}_2$	−1.22	0.48		$\hat{c}_2$	1.17	0.25
Colonization	$\hat{b}_{1,1}$	1.43	0.59	Detection	$\hat{d}_{1,1}$	−2.35	0.20
	$\hat{b}_{1,2}$	−8.17	3.13		$\hat{d}_{1,2}$	−10.19	2.00
	$\hat{b}_{2,1}$	2.33	0.47		$\hat{d}_{2,1}$	−1.80	0.09
	$\hat{b}_{2,2}$	−4.23	0.79		$\hat{d}_{2,2}$	−3.28	0.43
	$\hat{b}_{3,1}$	2.30	0.50		$\hat{d}_{3,1}$	−1.47	0.07
	$\hat{b}_{3,2}$	−2.11	0.36		$\hat{d}_{3,2}$	−2.18	0.24
	$\hat{b}_{4,1}$	0.67	0.48		$\hat{d}_{4,1}$	−1.43	0.04
	$\hat{b}_{4,2}$	−0.63	0.30		$\hat{d}_{4,2}$	−0.84	0.06
	$\hat{b}_{5,1}$	0.54	0.53		$\hat{d}_{5,1}$	−1.91	0.04
	$\hat{b}_{5,2}$	−0.74	0.34		$\hat{d}_{5,2}$	−0.35	0.04
					$\hat{d}_{6,1}$	−2.09	0.05
					$\hat{d}_{6,2}$	−0.43	0.05
					$\hat{d}_3$	0.94	0.03

et al. (2006) to approximate a diffusion model, although other approaches might also be feasible and provide a more mechanistic model (e.g., to model colonization as a function of the number of neighboring routes where house finches are present).

It should be noted that in this example, distance has been used as a covariate to model changes in the house finch's range, although other covariates could be used instead or in addition. Latitude and elevation might be especially relevant covariates for questions related to changes in species distributions caused by climate change.

9.6 Study Design

Good-quality data are the key to being able to make robust conclusions. This is no less the case with occupancy models as in any other area. For the two-category models, MacKenzie and Royle (2005) and MacKenzie et al. (2006)

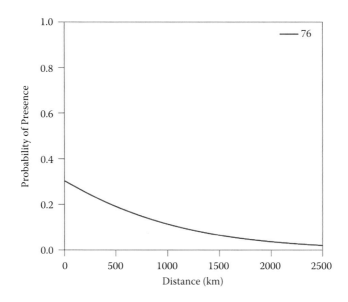

FIGURE 9.1
Estimated probability of house finch being present on a Breeding Bird Survey route in 1976 as
a function of distance from Long Island.

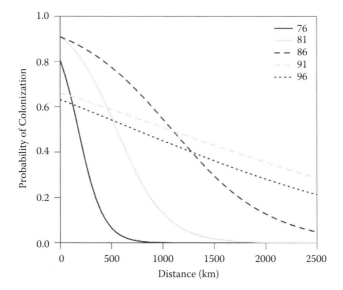

FIGURE 9.2
Estimated probability of house finch colonizing a Breeding Bird Survey route between successive 5-year periods as a function of distance from Long Island. The colonization probability is
indexed in terms of the beginning period; that is, "76" relates to the probability of colonization
occurring between 1976 and 1981.

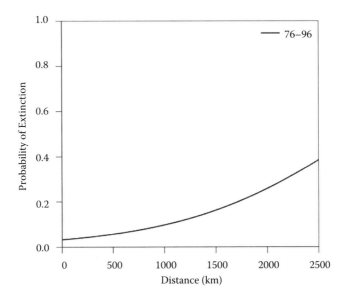

FIGURE 9.3
Estimated probability of house finch becoming locally extinct at a Breeding Bird Survey route between successive 5-year periods as a function of distance from Long Island.

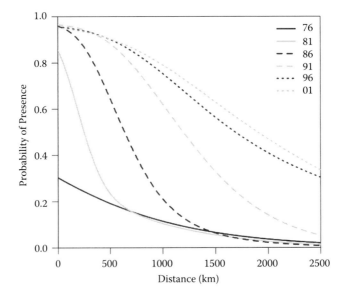

FIGURE 9.4
Estimated probability of house finch being present on a Breeding Bird Survey route in 1976–2001 as a function of distance from Long Island. Estimates are derived from occupancy in 1976 and colonization and extinction probabilities.

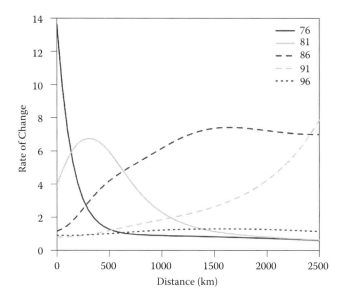

FIGURE 9.5

Estimated rate of change in house finch being present on a Breeding Bird Survey route as a function of distance from Long Island. Peaks indicate the distance at which the fastest rates of change occurred. Rate of change is indexed in terms of the beginning period; that is, "76" relates to the rate of change between 1976 and 1981.

provided guidance on study design issues, including definition of a sampling unit, what should be regarded as a season, and the required number of repeat surveys per season. The main points are highlighted next, as well as some updated thinking on these various issues. One main aspect of study design for occupancy models, with subtleties that can initially be difficult for some to fully grasp, is that no longer are we trying to sample individuals of the target species (whether they be plants or animals); we are sampling landscapes (or seascapes, riverscapes, etc.). The occupancy category at a sampling unit in the landscape is just a characteristic of that unit much like elevation, latitude, or vegetation type. Finding individuals of the target species within that unit is only one way to attempt to determine the occupancy category, and the primary intent of the field methods should not, necessarily, be to maximize the number of individuals captured.

9.6.1 Defining the Region of Interest

It is critically important to clearly define the region of interest to which the results are intended to apply as this defines the statistical population from which sampling units (however they are to be defined) are selected for surveying. Expanding or contracting the area of interest might create issues with the interpretation of results or lead to misleading conclusions about the changed region. Importantly, the conclusions are always specific to the statistical population from which the sample was drawn; hence, changing the population might change the results.

When defining the region of interest, some consideration should be given to whether there are certain areas that might be regarded as nonhabitat (i.e., no chance of the species being present there). If these areas are not of interest, then they could be excluded and no sampling effort be allocated to them. However, for longer-term studies or monitoring programs, if those areas might potentially become habitat or become otherwise occupied in the future, then some sampling effort should be devoted to those areas from the outset such that future changes can be identified. Actual surveys for the species might not have to be conducted from the outset, but those types of places should be included as part of the overall sampling frame. Similarly, if the intent is to identify changes in a species range over time, areas that are outside the current range should be included in the sampling frame to identify the species arriving at those locations. Absences should not be regarded as no information as they define which areas are outside the species current range.

9.6.2 Defining a Sampling Unit

As noted previously, the sampling unit is the item of interest for which the occupancy category is to be determined. For example, at this scale the intent might be simply to identify whether the species is present or absent. Depending on the context, the units may be naturally (e.g., ponds or forest patches) or arbitrarily (e.g., grid cells) defined, and it is the collection of all such units that defines the population of interest.

What would be an appropriate definition will depend on the exact application and the primary objective of the study. When interest is in a discrete set of naturally occurring units, such as ponds, then the choice of sampling unit will often be relatively straightforward. However, one additional consideration is whether the units themselves will be dynamic through time. For example, in a wetland complex or area of vernal pools, what may be considered a single pond in one year might be divided into multiple ponds the next or vice versa. Thus, the very population of interest (in a statistical sense) is changing each year, making it difficult to interpret any apparent changes in the occupancy metric. In such situations, an arbitrarily defined sampling unit might be more appropriate, with an initial consideration each year being whether there is currently water, for example, in the unit.

Where there is no natural definition, careful consideration should be given to the spatial extent to which an occupancy state is believed to apply. The spatial extent of a sampling unit is often implicitly assumed, and by considering such issues prior to collecting the data, these assumptions are being made explicit and can help formalize the sampling framework. For example, when a species has been detected at a location, it is unlikely that a dot on a map will be interpreted as the species is present only at that exact location; most would assume that the detection is indicative that the species is present in some larger area. The area over which the detection is believed to be indicative of presence would define the sampling unit.

Consideration of the home range, or territory, size of an individual of the target species has also been suggested as a means of defining a sampling unit, particularly for studies in which occupancy is interpreted as a surrogate for abundance. This might be a reasonable approach where individuals tend to occupy relatively well-defined natural features (e.g., drainages or habitat patches) but might not be that useful when sampling units (e.g., grid cells) are of an arbitrary size in a more contiguous landscape. For example, just because the grid cell has been defined at the approximate size of an individual's home range, it is unlikely that the home range will fall fully within a single grid cell because the species has no knowledge of the grid lines placed on a map; hence, a single individual might occupy multiple cells. Therefore, attempting to draw inferences about abundance from an occupancy metric might be misleading in this situation. In fact, generally speaking, there are relatively few cases for which studies can be suitably designed such that occupancy metrics could be meaningfully interpreted as something close to abundance.

It should be remembered that different definitions of a sampling unit may be suitable for different studies with different objectives even when working on the same species in the same general area. There are no hard-and-fast rules that will apply in most situations.

9.6.3 Defining a Sampling Season

Much like the definition of a sampling unit, sampling seasons have often been defined implicitly and encouraging such assumptions to be explicitly stated should improve the rigor of the study. Recall that the sampling season is the time period when the occupancy state of each sampling unit in the population is assumed to be static, or at least not changing in a nonrandom manner. This assumption is required so that the true occupancy state is observable for each survey within the season (i.e., that the units are closed to change in the occupancy state). Whenever the results of a single survey for a species are extrapolated beyond the actual survey period, then a similar assumption is being made. For example, if the detection of a species in a 5-minute point count is regarded as evidence the species is present at a site for the next 2 weeks (say), then that implies the occupancy status of the site is closed to changes for the next 2 weeks (i.e., there is a 2-week sampling season).

The sampling season does not have to correspond to a naturally occurring season such as a climatic or breeding season (although it may). It is the time period for which the occupancy status of a unit is applicable for a particular snapshot of the population. Therefore, the objective of the study or monitoring program should be the primary determinant of the sampling season.

9.6.4 Repeat Surveys

Repeat surveys provide the information that enables the occupancy status of the units to be disentangled (probabilistically) from imperfect detection. Although it is mathematically possible to separate the two with only a single survey (e.g., see Lele et al., 2012), it does require a number of potentially restrictive assumptions, including that occupancy and detection probabilities are functions of continuous covariates. However, the resulting estimates are not uniquely identifiable. Swapping the covariates for occupancy and detection provides the exact same regression coefficient estimates, although now the inferences about occupancy and detection are swapped. Therefore, the results are extremely model dependent.

The real value of repeat surveys is not that they enable the separation of occupancy and detection probabilities, but that they increase the probability of the true occupancy state being observed at least once within the sampling season. That is, the true occupancy state is more likely to be eventually observed during the defined season with additional repeat surveys; hence, results will be more robust and less model dependent. In the case with two occupancy states (e.g., species presence and absence), MacKenzie and Royle (2005) showed that sampling more sites might be inefficient if there is an insufficient number of repeat surveys (see the next section).

Repeat surveys do not necessarily imply distinct revisits to each site. MacKenzie et al. (2006) provide some detail of different ways in which the repeat survey information can be collected, but see also the work of MacKenzie and Royle (2005), Kendall and White (2009), and Guillera-Arroita et al. (2010). Some options include (1) multiple observers; (2) a single observer conducting multiple surveys in a single visit; and (3) spatial replication (e.g., multiple small plots within a larger sampling unit). Essentially, a survey is a single opportunity during which it is possible to detect the species. There are, however, trade-offs in terms of how the repeat surveys are conducted with respect to the effective sampling season length. For example, nominally the sampling season might be declared as a 2-week period, but if all surveys of a site are conducted on a single day, then the effective sampling season is also a single day, which might have important implications for how occupancy should be interpreted. The main considerations with respect to conducting repeat surveys is how they may have an impact on the main assumptions of the modeling.

9.6.5 Allocation of Effort

Given the need for repeat surveys, there are clearly competing requirements between the number of sampling units that should be surveyed versus the number of repeat surveys per unit, particularly when the total level of effort available is fixed and limited. Although little has been done to date on allocation of effort for applications with multiple occupancy categories, as noted previously, it has been shown for the two-category case that there is an optimal number of repeat surveys, as discussed by MacKenzie and Royle (2005), MacKenzie et al. (2006), and Guillera-Arroita et al. (2010). Even though the optimal number depends on how optimality is defined (Guillera-Arroita et al., 2010), from the perspective of trying to minimize the SE of the estimated occupancy probability for a given level of effort, or minimize effort to obtain a target SE, MacKenzie and Royle (2005) found that the optimal number of repeat surveys tended to result in the probability of detecting the species at least once during the repeat surveys being in the range 0.85–0.95. That is, the best estimate is obtained when the survey effort at each unit is of sufficient intensity that the presence or absence of the species is almost confirmed, at which point analyses that do and do not account for detection would likely provide similar results. This again suggests that the real benefit of repeat surveys is not that they enable the separation of the sampling (detection) and biological (occupancy) processes, but that they result in a higher-quality data set with less potential for false absences.

MacKenzie and Royle (2005) found that the optimal number of surveys depends on the per survey detection and occupancy probability (Table 9.2). The suggested number of repeat surveys is optimal in the sense that, for that same level of total effort, conducting fewer repeat surveys (therefore surveying more units) or more repeat surveys (hence fewer units) results in a less precise estimate of occupancy (e.g., as shown in Table 9.3).

The idea that there is an optimal number of repeat surveys that should be conducted each season also applies for multiseason studies. The best precision on estimates of colonization and extinction probabilities for a given level of effort will be achieved when the optimal number of surveys are conducted. On reflection, this is reasonable because, by reducing the possibility of a false absence within any particular season to a relatively low level, there

TABLE 9.2

Optimal Number of Repeat Surveys to Conduct per Sampling Unit for Various Levels of Detection p and Occupancy ψ

	ψ								
p	0.1	0.2	0.3	0.4	0.5	0.6	0.7	0.8	0.9
0.1	14	15	16	17	18	20	23	26	34
0.2	7	7	8	8	9	10	11	13	16
0.3	5	5	5	5	6	6	7	8	10
0.4	3	4	4	4	4	5	5	6	7
0.5	3	3	3	3	3	3	4	4	5
0.6	2	2	2	2	3	3	3	3	4
0.7	2	2	2	2	2	2	2	3	3
0.8	2	2	2	2	2	2	2	2	2
0.9	2	2	2	2	2	2	2	2	2

Source: MacKenzie, D.I., and Royle, J.A. (2005). *Journal of Applied Ecology* 42: 1105–1114.

TABLE 9.3

Expected Standard Error for Estimated Occupancy When $\psi = 0.4$ and $p = 0.3$ for Different Total Number of Surveys and Number of Surveys per Unit

Total Surveys	Surveys per Unit							
	2	3	4	5	6	7	8	9
100	0.22	0.16	0.14	0.14	0.14	0.14	0.15	0.15
200	0.16	0.12	0.10	0.10	0.10	0.10	0.10	0.11
500	0.10	0.07	0.06	0.06	0.06	0.06	0.07	0.07
800	0.08	0.06	0.05	0.05	0.05	0.05	0.05	0.05
1100	0.07	0.05	0.04	0.04	0.04	0.04	0.04	0.05
1400	0.06	0.04	0.04	0.04	0.04	0.04	0.04	0.04
1700	0.05	0.04	0.04	0.03	0.03	0.04	0.04	0.04
2000	0.05	0.04	0.03	0.03	0.03	0.03	0.03	0.03

Note: The number of units decreases as number of units increases.

is going to be less uncertainty about whether the species has truly colonized or gone locally extinct from a sampling unit.

9.7 Discussion

The set of methods that have been described are potentially applicable in any situation if questions of interest can be phrased in terms of species presence and absence or categories of occupancy. These are clearly not the only options but are certainly a useful addition to the toolbox of methods, particularly when it is likely that the true occupancy category at a sampling unit might be misclassified (e.g., by false absences). The ability to account for imperfect detection, which is typical in most field situations, is a clear improvement over methods that do not because imperfect detection that is unaccounted for will result in biased estimates and possibly misleading conclusions.

These approaches are relevant not only for data analysis but also can be used as the basis for making predictions about how species' distributions might change in the future. Once parameter values associated with the state transition probabilities have been estimated, these can be used to extrapolate forward from the existing situation 5, 10, 20, or 50 years through repeated application of the transition probability matrix. What-if scenarios can also be considered by changing the estimated values by a certain amount to identify how sensitive the future predictions might be to the estimated parameters.

There is a range of options for software to apply the techniques described in this chapter. The PRESENCE program is purpose-designed Windows-based freeware for implementing these types of methods, and the MARK program is alternative Windows-based freeware that can also be used for applying a subset of the models that are available in PRESENCE. Both options have also been successfully run on Mac or Linux systems using emulators. There is also an R package UNMARKED available that again includes the ability to fit a subset of the methods included in PRESENCE (as well as some additional models that are not in PRESENCE). Finally, OpenBUGS and JAGS are general statistical software that can be used to apply these methods within a Bayesian framework.

Although there has been rapid development of these methods since the early 2000s, there is still a lot to be learned about their performance under realistic field conditions to enable the efficient use of valuable logistical resources. Further refinements of the methods are sure to progress as the advantages and limitations of these techniques are explored.

10

Sampling Designs for Environmental Monitoring

Trent McDonald

10.1 Introduction

The statistical part of environmental monitoring designs contains three components: the spatial design, the temporal design, and the site design. The spatial design dictates where sample sites are located in the study area, the temporal design dictates when sample sites are visited, and the site design dictates what is measured at a particular site and how. These design components are largely, but not completely, independent. For example, methods in site design can substantially influence sample size and the population definition which in turn influence the spatial and temporal designs. However, the site design may simply require selection of a point as an anchor for field measurements, and the spatial design can select this point without knowledge of the specifics of the field measurements. Likewise, the temporal design can dictate when points are revisited without knowing their location or the specifics of field measurements. It is therefore useful and relatively easy to study general-purpose spatial designs without knowledge of the temporal or site designs. This is the approach taken by this chapter.

When designing the spatial component of an environmental monitoring study, the overriding purpose should be to select a sample of geographic locations that will eventually allow analyses that satisfy the study's objectives. In most cases, these objectives include making valid scientific inferences to all parts of the study area. In other words, the results of the final analysis (e.g., point estimates, confidence intervals, distributional statements, etc.) should be accurate and apply to the intended population.

This chapter considers key characteristics of spatial designs that serve these inferential purposes. It introduces some general-purpose sampling algorithms particularly suited for ecological monitoring that allow, when properly implemented, inferences to all parts of a defined study area. Although not exhaustive, the designs discussed here are appropriate and useful for a wide range of real-world, large-scale, long-term monitoring studies, as well as real-world, small-scale, or short-term studies. These designs, when combined and implemented at two or more nested levels, provide a great deal of flexibility.

The chapter is organized as follows: First, to provide background and context, a general discussion of design characteristics is given. In particular, the differences between scientific and nonscientific survey designs, as well as the differences between research and monitoring studies, are discussed. The differences between research and monitoring studies lead to sets of spatial sampling designs that are typically more appropriate for one type of study or the other. Next, spatial sampling algorithms appropriate for each type of study are described in separate sections, followed by some final comments and a summary.

Because the spatial design is concerned with geographic space, the terminology surrounding spatial designs is slightly different than classical finite-population sampling. Translation of a few classical finite-population terms is provided in Table 10.1.

TABLE 10.1

Traditional Finite-Population Sampling Terms and Their Equivalent Spatial Design Terms

Traditional Finite-Population Terms		Spatial Design Terms	
Term	**Definition**	**Term**	**Definition**
Sample unit	The smallest entity on which data will be collected	Sample site	A special case of sample unit that is tied to a geographic location; can be a point or a polygon
Population	A collection of sample units about which inference is sought	Study area	A collection of sample sites about which inference is sought; generally, a polygon or collection of polygons
Sample	A subset of sample units in the population	Spatial sample	A subset of sample sites in the study area; generally, a set of points or polygons in the study area; if polygons, the sum of their areas is necessarily less than total size of the study area

10.2 Design Characteristics

This section contains short and informal discussions about a few design characteristics that are important for practitioners. The section defines scientific surveys, gives a few examples of nonscientific surveys, and discusses the difference between research and monitoring studies.

10.2.1 Scientific Designs

A working and practical definition of scientific designs involves probability samples, which in turn depend on the definition of inclusion probabilities. Thus, inclusion probabilities (specifically, the first-order inclusion probabilities) are defined first, followed by probability samples, followed by the practical definition of scientific design.

The first-order inclusion probability (or sometimes just inclusion probability) is simply the probability of including any particular site in the sample. Technically, inclusion probabilities are strictly greater than 0 and less than or equal to 1. Inclusion probabilities can either be constant across sites or vary site to site. When inclusion probabilities vary, they commonly depend on strata membership or geographic location or are proportional to an external continuous valued variable associated with the site (such as elevation, annual rainfall, distance from the coast, etc.).

To make this definition concrete, consider the following example: Consider a study area partitioned into 100 quadrats (squares) and assume a researcher wishes to draw a simple random sample of 3 quadrats from these 100. A total of

$$\binom{100}{3} = \frac{100 \times 99 \times 98}{3 \times 2 \times 1} = 161{,}700$$

possible samples exists, and any particular quadrat is present in

$$\binom{99}{2} = \frac{99 \times 98}{2 \times 1} = 4{,}851$$

of them. The probability of including a quadrat is therefore $4{,}851/161{,}700 = 0.03$, as expected. If, instead of a simple random sample, the researcher ordered the list of quadrats by location, flipped a coin for each quadrat, and included the first three associated with heads, the inclusion probability of any particular quadrat would depend on its place in the list and the geometric distribution.

A working definition of a probability sample is a sample drawn in a way that the first-order inclusion probability of all sites can be known exactly. This definition admits those samples for which the first-order inclusion probabilities are not immediately known but could be calculated for all sites in the sample. In practice, the inclusion probabilities of sites not selected in the sample are not needed, but they need to be knowable if the sample is to meet the definition of a probability sample. To produce knowable inclusion probabilities, the algorithm selecting the sample must be repeatable and involve some sort of stochastic component.

A working definition of scientific design is then any sampling algorithm that results in a probability sample. In this view, a probability sample and scientific design are synonymous. The term *scientific design* is used in what follows because it is common and is often more natural in communications.

In general, nonscientific designs are simply those that do not include a probability sampling component. Under this definition, the universe of nonscientific designs is vast and varied. Generally, nonscientific designs either are not repeatable or do not involve some sort of random component.

The perils of nonscientific designs are well known (McDonald, 2003; Olsen et al., 1999; Edwards, 1998). Yet, nonscientific designs remain common and overtly attractive. Nonscientific designs are attractive primarily because they are easier, quicker, and less expensive to implement than scientific designs. To illustrate the point that nonscientific designs are common, three types are described next.

10.2.2 Judgment Samples

Judgment samples are those for which researchers, usually familiar with the resource, make a judgment or decision regarding the best places to locate sample sites. To justify this, researchers may represent that they placed sample sites in places that are representative, where the most change is anticipated, in critical habitat, where the impact will be, or where interesting things will happen. In doing so, the researchers are making judgments or assumptions about relationships among the study's parameters that may be true or false. In limited cases, when a research study's objectives are extremely narrow or if the study is a true pilot study, judgment samples may be acceptable. But, even in these cases, the assumptions and judgments made during site selection need to be acknowledged and tested. The real danger of judgment samples is that the researcher's assumptions are false, or only partly true, and the study results are biased.

10.2.3 Haphazard Samples

Haphazard samples are those collected without a defined or quantifiable sampling plan. Under haphazard sampling, the placement of sample sites is not usually planned in advance. Many times, haphazard samples are placed and taken where and when field crews have extra time after performing other duties. In many cases, field crews are responsible for determining sample site placement while they are in the field, and unless crews are aware of the pitfalls, sample sites are often placed wherever the crew happens to be. When haphazard spatial samples are used, meaningful general-purpose analyses are difficult. Haphazard and unplanned sample sites may provide interesting observations but should not be considered part of scientific monitoring plans.

10.2.4 Convenience Samples

An extremely popular form of nonscientific designs are convenience samples. Convenience samples are those that include sites because they are easy and inexpensive to reach. Common convenient sample sites include those located close to research facilities, roads, or other access ways (such as trails). Because budgets are always limited, convenience samples are a tempting and popular way of reducing costs. The danger with convenient samples is bias. Convenient sites, or easy-to-access sites, may or may not represent the entire population. Sites that are easy to access can and often are changed by the fact that they are easy to access and therefore are different from more remote sites. Furthermore, to justify that the convenient sites are similar to remote sites, it is necessary to place at least some sites in remote areas, and at that point, it would be easier and more defensible to select a (perhaps stratified) probability sample from both areas.

Convenience samples can employ probability samples of the areas that are easy to access yet remain a convenience sample if the study area is not redefined to correspond to the region that was actually sampled. In other words, drawing a probability sample from easy-to-access areas is perfectly valid unless inference to a larger study area is desired. For example, drawing a probability sample near roads is perfectly acceptable as long as it is clear that estimates apply to regions of the study area near roads, where *near* must be defined. It is a mistake, however, to represent estimates constructed from a near-road probability sample as applying to regions far from roads as well.

10.3 Monitoring versus Research

In addition to the differences between scientific and nonscientific designs, there are differences between monitoring and research studies that have an impact on the spatial sampling designs of both. The designs described later in this chapter are classified as generally appropriate for either research or monitoring, and this section defines both types.

The primary defining characteristic of research studies is that they answer a specific question or estimate a specific parameter. Research studies generally focus estimation efforts on one or two target variables measured over a relatively brief period of time (say 1 to 5 years). They are convenient for graduate students because objectives are usually focused so that the study can be performed during a graduate student's normal tenure. Research studies usually address an immediate problem or crisis or inform an impending management decision. When that problem, crisis, or decision is resolved, motivation to continue the study drops. For this reason, a spatial design that provides the highest possible statistical precision to estimate a single parameter in the shortest amount of time is usually paramount.

In contrast, a key characteristic of most monitoring studies is that they are not designed to answer a specific question. The objectives of monitoring studies are usually summarized as watching an environmental resource. For example, objective statements for monitoring commonly contain phrases like "to estimate current status and detect trends." As such, and especially when compared to research studies, the objectives of monitoring studies can seem vague. Monitoring studies tend to be long in duration, typically 10 to 30 years, because annual variation of parameters is usually large and trend detection requires an extended period. Monitoring studies also tend to be large scale in their geographic scope, where large means that the study incurs significant travel between sample sites. Monitoring studies also tend to be funded by agencies with management authority over the resource.

Spatial designs used in monitoring studies generally cannot focus on a single variable or objective, while designs used in research studies can be optimized to provide the highest possible precision for a single parameter. Real-world monitoring studies almost always have multiple objectives, and it is difficult to optimize the placement of locations for estimation of multiple parameters. For a monitoring study to survive one to three decades, its spatial design needs to be robust enough to provide high quality data on multiple parameters. It must be easy to implement and maintain, and it must be able to provide data on unforeseen issues when they arise.

Unfortunately, much of the statistical literature on spatial design is relevant only when interest lies in a single parameter. Experimental design books such as those by Steel and Torrie (1980) and Quinn and Keough (2002) discuss techniques like blocking and nesting as ways to maximize the ability to estimate a treatment affect. Stratification, discussed in Chapter 2, is usually thought of as a technique to improve precision, but it generally works only for a single parameter. Maximum entropy (Shewry and Wynn, 1987; Sebastiani and Wynn, 2000), spatial prediction (Müller, 2007), and Bayesian sampling methods (Chaloner and Verdinelli, 1995) are designed to locate sites in a way that maximizes a criterion (such as information gain). But, the criteria used by these designs are functions of a single variable; thus, these procedures optimize for one parameter at a time. Focusing on one parameter is generally undesirable in monitoring studies because near-optimal designs for one parameter are suboptimal for other variables unless spatial correlation is strong.

10.4 Spatial Designs

The main part of the remainder of this chapter consists of two sections. The research designs section relates four common designs that are generally more appropriate for research studies than monitoring studies, although the designs in this section can be applied to monitoring studies under certain circumstances (e.g., when there are only a few monitored variables). These designs, listed in Table 10.2, consist of simple random sampling, two-stage sampling, stratified sampling, and cluster sampling. The monitoring designs section describes three designs that are generally more appropriate for monitoring studies. These designs are the two-dimensional systematic sample, the one-dimensional general random sample (GRS), and d-dimensional balanced acceptance sample (BAS). An important characteristic of these designs is that they ensure a high degree of spatial coverage. R code to draw the more complex samples (i.e., GRS and BAS) is available on the book's web site (https://sites.google.com/a/west-inc.com/introduction-to-ecological-sampling-supplementary-materials/).

10.4.1 Research Designs

This section contains brief descriptions of four common spatial sampling designs that are generally appropriate for research studies. As noted, a long-term study's interest may lie in a single variable, and in this case, the designs of this section could be applied in a monitoring setting. Also, certain of the sampling plans (i.e., simple random, systematic, and GRS) can be applied at different stages of a larger design. For example, it is possible to draw a systematic sample of large collections of sampling units at one level, followed by a simple random sample of units from within the large collections. There is more about these finite-population designs in Chapter 2 of this book, and they are covered extensively in the books by Cochran (1977), Scheaffer et al. (1979), Särndal et al. (1992), Lohr (2010), and Müller (2007).

Except for simple random sampling, a significant amount of a priori information is required about the target variables and the study area before a sample can be drawn. For instance, to implement a stratified design, regions of the study area must be classified into categories. This categorization requires knowing or estimating an auxiliary variable on which strata are defined. If a maximum entropy design is to be implemented, some information about the magnitude and structure of spatial covariance must be known. If a cluster design is to be implemented, the size and configuration of clusters must be known throughout the study area.

10.4.1.1 Simple Random Sampling

Simple random samples of a geographic study area are drawn by first bounding the study area with a rectangular box to delineate the horizontal x and

TABLE 10.2

Common Spatial Designs Suitable for Research and Monitoring Studies, Along with References and a Brief Description or Comment

Spatial Design	References	Description/Comment
Research		
Simple random	Cochran (1977, Chap. 2) Scheaffer et al. (1979, Chap. 4) Särndal et al. (1992, Chap. 3) Lohr (2010, Chap. 2)	Sites selected completely at random. Covered by many sampling texts and articles.
Two stage	Cochran (1977, Chap. 11) Scheaffer et al. (1979, Chap. 9) Särndal et al. (1992, Chap. 4) Lohr (2010, Chap. 6)	Large or primary sites (polygons) selected first. Small or secondary sites within the selected primary sites are sampled second. In general, the design and sample size can vary by stage.
Stratified	Cochran (1977, Chap. 5, 5A) Scheaffer et al. (1979, Chap. 5) Särndal et al. (1992, Chap. 3) Lohr (2010, Chap. 3)	A special case of two-stage sampling in which strata define primary sites, and all primary sites are sampled (census at stage 1). A special case is the one-per-strata sampling (Breidt, 1995). In general, design and sample size can vary by strata.
Cluster	Cochran (1977, Chap. 9, 9A) Scheaffer et al. (1979, Chap. 9) Särndal et al. (1992, Chap. 4) Lohr (2010, Chap. 5)	A special case of two-stage sampling that occurs when all secondary sites within a primary site are sampled (census at stage 2).
Monitoring		
Systematic	Cochran (1977, Chap. 8), Scheaffer et al. (1979, Chap. 8) Särndal et al. (1992, Chap. 3) Lohr (2010, Section 2.7)	Also, grid sampling. Suitable for two-dimensional resources. The geometry of the grid is usually square (square cells) or triangular (hexagonal cells).
General random sample (GRS)	This chapter	Suitable for one-dimensional resources. Produces equal probability or unequal probability samples, ordered or unordered samples, simple random and systematic samples of fixed size.
Balanced acceptance sampling (BAS)	Robertson et al. (2013)	Suitable for n-dimensional resources. Produces equal probability or unequal probability spatially balanced samples in n dimensions.

Note: See corresponding section of the text and references for a more thorough description.

vertical y range of coordinates. Random site coordinates are then generated one at a time as random coordinate pairs within the bounding box. Random coordinate pairs are generated by choosing a random deviate from a uniform distribution for the x coordinate, and an (independent) random deviate from a uniform distribution for the y coordinate. When generated, the random point (x,y) will fall somewhere within the study area's bounding box but not necessarily inside the study area. If the random point (x,y) falls outside the study area, it is discarded and another point generated. The process of generating points inside the bounding box is repeated until the desired sample size in the study area is achieved. An example of a simple random sample from a rectangular study area is shown in Figure 10.1.

Simple random samples provide an objective assessment of the study area. If a particular sample site is inaccessible or unsuitable, additional sites are easily generated and the sample will retain its basic statistical properties. Simple random sampling is the only design listed in this section that does not utilize *a priori* information on target variables and therefore can be easily and rapidly implemented in most studies. Simple random sampling, however, suffers from the fact that it does not guarantee uniform coverage inside the study area. It is possible for clumps of sample locations and large nonsampled areas to appear. For example, the sample shown in Figure 10.1 contains a conspicuous vertical strip approximately one-third of the way from left to right that is devoid of sample sites.

Assuming that spatial correlation exists in the study area, these clumps and nonsampled areas leave some regions of the study area overrepresented and other regions underrepresented. Over- or underrepresentation of particular areas is not a problem statistically but is inefficient. It should be noted that if no spatial correlation exists, a simple random sample may as well be

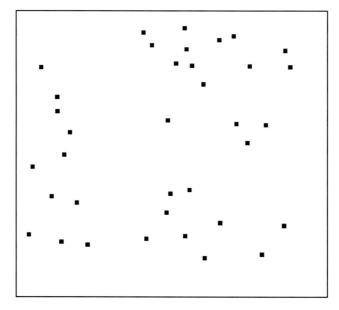

FIGURE 10.1
Example of a simple random sample of size 36 from a square study area.

drawn because all areas are equally represented from a statistical point of view regardless of sample site locations.

10.4.1.2 Two-Stage Sampling

Two-stage designs (Cochran, 1977, Chapter 11; Scheaffer et al., 1979, Chapter 9; Särndal et al., 1992, Chapter 4; Lohr, 2010, Chapter 6) contain two nested levels of sampling. Two-stage sampling is implemented when sites can be naturally grouped into large collections and it is relatively easy to select the large collections. The large collections of sample sites are called primary sample units and are defined to exist at level 1 of the study. When geographic areas are sampled, the primary units are usually large polygons such as watersheds, sections, counties, states, and the like. In general, primary units could be any structure containing multiple sites or sample units. Under this design, primary units are selected using one of the basic spatial designs, such as simple random sampling, systematic sampling, GRS, or BAS.

Once primary sample units are defined and selected via some design, the actual locations for sample sites are selected from within the chosen primary units. In this setting, sample sites are called secondary sample units and are said to exist at level 2 of the study. Secondary units are selected from each primary unit using one of the basic spatial samples, for instance simple random, systematic, GRS, or BAS. It is not necessary for the sampling design or sample size to be consistent across primary units but usually it is. An example two-stage sample appears in Figure 10.2.

It is possible to implement a design with three or more levels when sample units naturally nest within one another. For example, when sampling the entire Unites States, it may make sense to select states, then townships within states, then sections within townships, then sites within townships.

Two-stage designs with less than a census of primary units at level 1 are relatively rare in ecological studies. More commonly, every primary unit at stage 1 is selected, and all sampling is done at stage 2. In this case, the design is said to be stratified (see next section). However, if there are many large primary units, selecting a spatially balanced sample of primary units followed by a spatially balanced sample of secondary units may be a good design for either research or monitoring purposes.

10.4.1.3 Stratified Sampling

Stratified designs, as covered in Chapter 2, are special cases of two-stage designs that occur when all primary units are selected and a sample of secondary units is drawn from each. When all primary units are sampled, the primary units are renamed strata. Collectively, the strata partition the study area into mutually exclusive sets of secondary units (sites). An example stratified sample appears in Figure 10.3. Like two-stage sampling, sample size and even design can vary between strata.

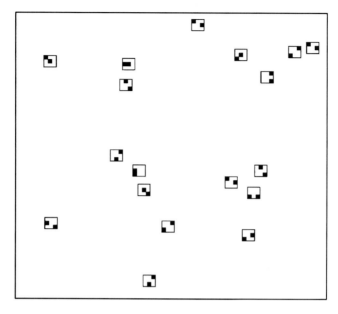

FIGURE 10.2
Example of a two-stage sample of 18 primary units and 2 secondary units from each, for a total of 36 sample sites. Each primary unit was defined to be a 3 × 3 block of 9 secondary units. Primary and secondary units were chosen by simple random sampling.

In Figure 10.3, a simple random sample of size 16 was drawn from each of three strata. Because the strata in this example differ in size, the density of sampled sites in each strata varies. When this happens, the sample is said to be nonproportionally allocated. If the density of sites is equal in all strata or is proportional to strata size, the sample is said to have proportional allocation. In Figure 10.3, the left-hand stratum is twice as large as the other two. Proportional allocation would have been achieved if sample sizes were 36/2 = 18 in the large strata and 36/4 = 9 in each of the two smaller strata. Proportional allocation can be important if poststratification of the sample is anticipated. Poststratification is easy under proportional allocation because all sites are included with equal probability. If allocation is not proportional, poststratification introduces unequal probabilities into the analysis, which yields a statistically valid but complicated analysis.

In ecological studies, strata are usually geographic regions of interest or are based on the perceived level of a target variable. For example, a study of fish populations in rivers might stratify based on elevation under the assumption that higher-elevation rivers contain more (or less) fish than lower-elevation rivers. Another fish study might attempt to classify stream segments into low-, medium-, or high-productivity strata and draw samples for each productivity level.

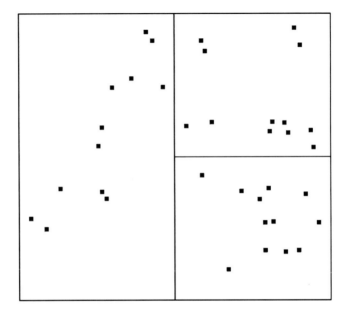

FIGURE 10.3
Example of a stratified sample of 36 sites from three strata. In each stratum, 12 sites were selected using the simple random design.

When constructing strata, it is best to base strata boundaries on factors that either do not change or change slowly. Examples of good strata boundaries are geographic (mountain ranges, elevation, etc.) or political boundaries (states, counties, etc.). Examples of poor strata boundaries are those based on habitat classifications or distances from roads. These criteria make poor strata boundaries because they have the potential to change rapidly.

10.4.1.4 Cluster Sampling

Cluster samples are special cases of two-stage samples that occur when all secondary units from selected primary units are selected. Like two-stage sampling, cluster sampling draws a sample of primary units at stage 1. At stage 2, however, a cluster design selects all secondary units within each selected primary. An example cluster sample is shown in Figure 10.4.

In ecological studies, cluster sampling can be useful when all sample units of a collection (cluster) are relatively easy to collect. For example, in some situations it may be relatively easy to capture an entire group of gregarious animals (e.g., wild horses, dolphins, etc.). In these cases, it is reasonable to define the group as a cluster and individuals within the group as the secondary units. Cluster sampling is not popular for sampling geographic locations, except for the special case of systematic sampling, unless the number of clusters sampled is high.

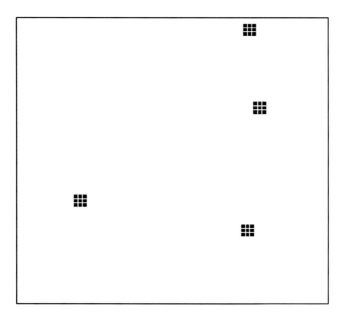

FIGURE 10.4
Example of a cluster sample of size 4. Each primary unit was defined to be a 3×3 block of 9 secondary units. A total of $9 \times 4 = 36$ sites were chosen.

10.4.2 Monitoring Designs

Although there are many spatial designs generally applicable to research studies, and even more methods for improving or optimizing the spatial design for a single parameter (Müller, 2007), there are relatively few general-purpose spatial designs that perform well in monitoring studies. The primary characteristic of the few designs that perform well in monitoring studies is that they ensure broad spatial coverage of sites (Kenkel et al., 1989; Schreuder et al., 1993; Nicholls, 1989; Munholland and Borkowski, 1996; Stevens and Olsen, 2004; Robertson et al., 2013). Ensuring broad spatial coverage works well in monitoring studies because spatial variation is often one of two large sources of variation, the other large source being temporal variation.

Of the designs that ensure good spatial coverage, three have the highest likelihood of satisfying the objectives of a monitoring study. Other designs can work, but they are usually more difficult to apply in a way that satisfies all objectives. The three sample designs generally suitable for monitoring studies are systematic, GRS, and BAS. A fourth design, generalized random tesselation stratified (GRTS) (Stevens and Olsen, 2004), has been implemented in many environmental studies, but the next generation of BAS is easier to implement than the GRTS design and has better spatial coverage properties (Robertson et al., 2013).

10.4.2.1 Systematic Sampling in Two Dimensions

Systematic samples, also called grid samples, are special cases of cluster sampling for which each cluster is spread over the entire study area (i.e., secondary units are not contiguous) and only one cluster is sampled. Systematic samples derive their name from the fact that sample sites are systematically placed. That is, after an initial random start, systematic samples choose secondary units in the same relative location within groups or blocks of units. The size and shape of each block are the design's step size (or grid spacing) and define the number of secondary sites chosen. In general, systematic samples do not yield a fixed number of sites per draw. That is, the clusters do not all contain the same number of sites unless the study area size is evenly divisible by the sample size (in all directions). Although varying sample size can pose problems logistically, it is known that varying cluster sizes do not pose any statistical problems (Särndal et al., 1992). This section discusses two-dimensional systematic samples. One-dimensional systematic samples are possible and should be implemented as GRS as discussed in the next section.

A systematic sample of a rectangular geographic region is shown in Figure 10.5. This has dashed lines to show boundaries of blocks of secondary units from which a single unit is selected. The width and height of these blocks are the horizontal and vertical step sizes. The width and height of blocks are determined by the sample size and the width and height of the

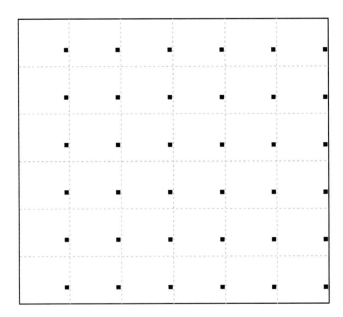

FIGURE 10.5
Example of a 6 × 6 systematic sample with random start. Dashed lines indicate blocks of secondary units (sites) from which a single unit is selected. The set of sites in the same relative position in each block constitutes a cluster.

study area. In the example of Figure 10.5, the desired total of 36 sites was arbitrarily arranged into 6 rows and 6 columns, and the number of sample units in the vertical and horizontal directions was evenly divisible by 6. A different configuration of 36 sites arranged in 3 rows of 12 appears in Figure 10.6. Once block sizes are determined, a random location within the first block is generated. The site at that location is then selected and the remainder of the sample is filled with sites in the same relative position from other blocks.

If the study area is an irregular polygon, a systematic sample can be drawn by laying the (randomized) grid of sample sites over an outline of the study area and choosing grid points that fall inside. This process of intersecting a randomized grid and a nonrectangular study area produces unequal sample sizes because different numbers of sites fall in the study area depending on the randomization, but other than logistical difficulties, varying numbers of sample sites do not pose problems for analysis. Under systematic sampling, the inclusion of units is not independent. Given the identity of one unit in the sample, the identities of all other units in the sample are known.

Systematic samples are good designs for monitoring studies. They are useful for monitoring because the realized sample has good spatial balance. However, systematic sampling suffers because the realized sample size is usually variable, and it can be difficult to replace sample sites that are inaccessible or inappropriate. If a site is inaccessible or inappropriate and researchers wish to replace it with another, it is difficult to decide where

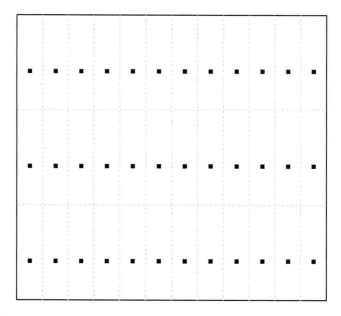

FIGURE 10.6
Example of a 3 × 12 systematic sample with random start. Dashed lines indicate blocks of secondary units (sites) from which a single unit is selected. The set of sites in the same relative position in each block constitutes a cluster.

the additional site should be placed. Replacement is difficult because any placement of fewer than a full grid reduces the spatial balance of the sample; however, the replacement of systematically chosen sites does not cause inferential problems any more than replacement of sites under other designs.

Spatial balance can also suffer if budget and time run out before the systematic sample is completed. If budget or time constraints preclude completion of a full systematic design, it is possible for large sections of the study area to go unsampled unless researchers purposefully order site visits to preserve the spatial balance under all realized sample sizes. Guarding against spatial imbalance under foreshortened field seasons means researchers cannot visit sites in a systematic order, and this reduces logistical efficiency. This difficulty in maintaining spatial balance under addition or subtraction of sites is not present in BAS samples, but BAS sites cannot be visited in an arbitrary order.

10.4.2.2 General Random Samples in One Dimension

GRSs allow multiple approaches to sampling one-dimensional resources. GRSs can be used to draw equal probability or unequal probability, ordered or unordered, simple random or systematic samples over one-dimensional resources. Samples drawn by the algorithm have a fixed size. Although GRS does not generalize easily to two dimensions, one-dimensional resources encompass a large number of natural resource types, such as streams, beaches, lake edges, forest edges, unordered frames, and so on. Both finite and infinite one-dimensional resources (such as points on a line) can be sampled using the GRS algorithm, assuming the infinite resources can be discretized into a finite list.

GRSs of size n are drawn as follows: Let a vector of auxiliary variables associated with each sample unit in the population be denoted $\mathbf{x}$. The vector $\mathbf{x}$ could be constant or could contain values such as the size of the unit, the distance from a particular location, or the anticipated level of a target variable. When $\mathbf{x}$ is constant, the design is equiprobable. When $\mathbf{x}$ is not constant, the sample is drawn with probabilities proportional to $\mathbf{x}$. Drawing the GRS involves first scaling the values in $\mathbf{x}$ to sum to n by dividing by their sum. If any scaled values are greater than 1, they are set equal to 1, and the remaining values are rescaled to sum to $n - k$, where k is the number of values greater than 1. A random start between 0 and 1 is then chosen and used as the starting point for a systematic sample (step size = 1) of units associated with the scaled values. A heuristic pictorial presentation of the GRS algorithm is given in Figure 10.7. R code for drawing a GRS is provided on the book's website (https://sites.google.com/a/west-inc.com/introduction-to-ecological-sampling-supplementary-materials/).

Note that the GRS algorithm is silent regarding the ordering of units in the population. If all elements in $\mathbf{x}$ are equal and the order of units in the population is randomized prior to selecting the GRS, the resulting sample

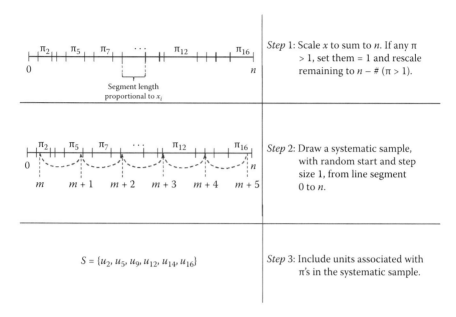

	Step 1: Scale x to sum to n. If any π > 1, set them = 1 and rescale remaining to $n - \#\,(\pi > 1)$.
	Step 2: Draw a systematic sample, with random start and step size 1, from line segment 0 to n.
$S = \{u_2, u_5, u_9, u_{12}, u_{14}, u_{16}\}$	*Step* 3: Include units associated with π's in the systematic sample.

FIGURE 10.7
A heuristic pictorial representation of drawing a general random sample (GRS) of size $n = 6$ from a population of size $N = 16$.

is a simple random sample. If all elements in **x** are equal but the order of units in the population is fixed, the resulting sample is a one-dimensional, fixed-size systematic sample. This type of systematic sample is appropriate when it is desirable to order units according to some auxiliary variable, such as the distance from a geographic location, elevation, easting, or northing. For example, it might be desirable to order stream segments by river mile (distance from the mouth) to ensure that sample sites are located in all parts of the river. If elements in **x** are not equal and the order of units is randomized, the resulting sample is a simple random sample with probability of inclusion proportional to **x**. If elements in **x** are not equal but the order of units is fixed, the resulting sample is a systematic sample with probability of inclusion proportional to **x**.

10.4.2.3 Inclusion Probabilities for GRS

At analysis time, it is important to know or at least estimate the properties of the sampling design under which data were collected. Principally, this involves computing or estimating the design's first-order and second-order inclusion probabilities. Third- and higher-order inclusion probabilities are usually ignored. First-order and second-order inclusion probabilities are important because they are used in the Horvitz–Thompson (Särndal et al., 1992) and other estimation techniques.

For the GRS algorithm, first-order inclusion probabilities are easy to obtain. These are the scaled interval lengths used in the final systematic sampling step. That is, the first-order GRS probability of unit i being selected is its associated scaled value of x_i. The second-order inclusion probabilities for a GRS are more difficult to obtain. In theory, second-order GRS inclusion probabilities are computable, but the computations required are time consuming. Consequently, second-order inclusion probabilities are usually approximated, either by a formula or by simulation. Stevens (1997) gave formulas for special cases of GRTS samples that can also be used to approximate second-order inclusion probabilities for GRS samples. Second-order inclusion probabilities in other cases will usually be approximated by simulation. The simulation needed to approximate second-order inclusion probabilities replicates the GRS algorithm and tallies the number of times each pair of units occurs in the sample. R code to implement this type of simulation appears on the book's web site.

10.4.2.4 Balanced Acceptance Samples

As the name implies, balanced acceptance samples (BAS) are a form of acceptance sampling that ensures spatial balance. In this section, the algorithm for drawing an equiprobable BAS sample in two dimensions is described and illustrated. BAS algorithms that sample n-dimensional resources and can incorporate variable inclusion probabilities are possible but are beyond the scope of this book. The description here is a condensed version of the description in Robertson et al. (2013).

10.4.2.4.1 Halton Sequence

The Halton sequence (Halton, 1960) is well known in mathematics as a pseudorandom number generator and as a method that evenly distributes points throughout a space. The Halton sequence serves sampling purposes by effectively mapping d-dimensional space to a one-dimensional sequence of numbers. The d-dimensional Halton sequence is, in fact, a combination of n one-dimensional sequences (each being van der Corput sequences), and these one-dimensional sequences are defined first.

The sequences that spread sample locations throughout a one-dimensional resource are constructed by choosing a number base, say, p ($p \geq 2$) and then reversing the base p representation of the integers 1, 2, and so on. For example, the base 3 representation of 10 is 101 (i.e., $1(3^2) + 0(3) + 1 = 10$). The reversed base p representation, often called the radical inverse, of interest here is

$$\varphi_3(10) = 0.101 = \frac{1}{3^1} + \frac{0}{3^2} + \frac{1}{3^3} = \frac{10}{27} = 0.3704.$$

The base 3 representation of 11 is 102, and the reversed base 3 representation of 11 is

$$\varphi_3(11) = 0.201 = \frac{2}{3} + \frac{0}{9} + \frac{1}{27} = \frac{19}{27} = 0.7037.$$

A property of these reversed base p sequences is that any contiguous subsequence is evenly distributed over the interval [0,1] (Wang and Hickernell, 2000).

A d-dimensional Halton sequence is simply d one-dimensional reversed base sequences, one for each dimension, but with one extra condition. The extra condition, which in fact ensures the spatial balance, is that bases of all one-dimensional sequences must be pairwise coprime. Pairwise coprime means all bases must be prime and unique (no repeats). Although any set of coprime bases will satisfy the definition of a Halton sequence, for the purposes of sampling two-dimensional space, the BAS algorithm chooses $p_1 = 2$ and $p_2 = 3$. For the purposes of sampling, stochasticity of the selected points is added to the Halton sequence by starting each one-dimensional sequence in a random place. A Halton sequence started in random placed in each dimension is called a randomized Halton sequence.

10.4.2.4.2 Equal Probability BAS Design

To draw an equiprobable BAS sample, first define a rectangular bounding box surrounding the study area. Generate a randomized Halton sequence inside the bounding box and take points, in order, until n locations in the study area are obtained. Points landing outside the study area are discarded. A realization of an equiprobable BAS design in the Canadian province of Alberta is shown in Figure 10.8. The points displayed in Figure 10.8 were drawn using the R package SDraw.

10.4.2.4.3 Unequal Probability BAS Design

The BAS design described in the previous section can be modified to draw unequal probability samples in continuous space. Drawing an unequal probability sample is accommodated by adding a dimension proportional to inclusion probabilities and implementing an acceptance sampling scheme. See the work of Robertson et al. (2013) for a description.

10.4.2.4.4 Inclusion Probabilities for BAS

Robertson et al. (2013) noted that the first- and second-order inclusion probabilities for regions in the study area can be computed by enumerating all possible randomized Halton sequences. Although enumerating all possible Halton sequences is not practical for large problems, it is computationally feasible in many cases, especially if the computations are processed in parallel. If it is not feasible to enumerate all randomized Halton sequences, first-

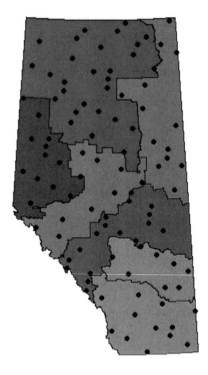

FIGURE 10.8
A BAS sample of size n = 100 in the Canadian province of Alberta. The BAS sample was drawn using the SDraw package in R.

and second-order inclusion probabilities can be approximated by simulation using code similar to that used for GRS samples (Robertson et al., 2013).

10.5 Summary

In this chapter, concepts important when designing a sampling scheme were discussed. Concepts included the definition of scientific and nonscientific designs. Scientific designs are those that draw probability samples; nonscientific designs are those that do not. It is suggested that environmental studies, and monitoring studies in particular, should always use scientific designs.

Another concept discussed was the distinction between research and monitoring studies. This distinction was made because the sampling design needs of both studies are generally different. In particular, the sampling design of research studies must usually provide the highest possible statistical precision for a particular parameter or hypothesis. The sampling design of monitoring studies must provide adequate data to address a wide range of anticipated and unanticipated estimation tasks over a long time frame. Four sampling designs generally appropriate for research studies were described: simple random, two-stage, stratified, and cluster sampling. Three spatial designs generally appropriate for monitoring studies were described: systematic sampling, GRS (for one-dimensional resources), and BAS designs.

11

Models for Trend Analysis

Timothy Robinson and Jennifer Brown

11.1 Introduction

Previous chapters of this book have focused on various strategies for sampling ecological populations to estimate parameters such as the population size or density. Often, environmental managers require information not only on the current status of a population but also on the changes in the population over time. For example, interest may be in monitoring a special area such as national parks (e.g., Fancy et al., 2009), monitoring for biodiversity (Nielsen et al., 2009), monitoring to detect the spread of invasive species (Byers et al., 2002), or monitoring of pollution or contamination levels (Wiener et al., 2012). Monitoring for changes over time helps discern whether management actions are needed for a vulnerable species or ecosystem and for assessing whether management actions have been effective. For example, the Convention on Biological Diversity, which aimed to reduce the rate of loss of biological diversity, referred to indicators to assess progress toward the 2010 target (http://www.biodiv.org/2010-target). Included in these indicators was monitoring the changing abundance and distribution of selected species, as well as measuring changes in the status of rare or threatened species.

In this chapter, we consider various methods of assessing change over time. We use a data set of lake contamination measured by mercury concentrations in fish (in ppm). Fish were collected by trawls at different locations around the lake. Fish tissue was minced and then assayed for mercury concentration. The trawls were conducted at 10 randomly chosen stations in the lake for 12 years. The feature of this design is that the same stations were repeatedly measured, and this in turn means that the structure of the observation data was clustered. The other feature of the data is that they are balanced; that is, there are equal numbers of observations ($n_i = 12$) for each of the 10 stations, giving a total of $n = 120$ observations in the data set.

One of the goals of the study was to determine if there was a significant trend in the contamination levels over the 12 years of surveying. Another goal was to assess the variability in contamination across the stations and to see if this variability was more, or less, than the variability in mercury concentration for each station. The first goal mentioned directly relates to determining changes in lake pollution, whereas the other goal relates to decisions about future allocation of sampling effort if the study is to continue. For example, if there were more sampling effort available, the decision would need to be made about whether to increase the frequency of observations per station or to increase the number of stations.

When repeated observations are made on each sampling unit, there are two ways to approach the data: through a unit analysis or through a pooled analysis. The unit analysis involves a separate analysis on each sampling unit (e.g., transect, quadrat, radio-collared animal, etc.). Results from the individual units that are summarized across the units can be used to make inferences about the population of all units. In our example, this would involve first looking at the data from each station separately. Pooled analyses generally involve mixed models (Myers et al., 2010). These models exploit the fact that the sampling units are a random sample from a larger population of units. The sampling units will vary from each other because of unmeasured factors, such as lake depth, water currents, and aquatic vegetation. For terrestrial studies, these factors may be forest composition, elevation, aspect, and other intrinsic factors related to population dynamics. These unmeasured factors are essentially rolled together into sampling unit variance components. In our example, it is easy to imagine that there will be variation in mercury measurements among stations because of small-scale variation in the lake bed structure and fish diversity.

Mixed models allow for inferences to be made on individual units as well as to the population of units as a whole. Although more complicated than the unit analyses, there are many advantages. For example, these models can easily accommodate missing data, such as when there are unequal numbers of observations on each sampling unit. Mixed models can also be applied to the analysis of data organized with multiple layers of nesting (e.g., monthly data recordings nested within seasons).

11.2 Basic Methods for Trend Analysis

The use of simple but effective graphics is an essential part of initial data exploration. Graphical displays of the data are useful for reporting on population changes because they can be an effective way to convey to others the main trends. An example is the index plot shown in Figure 11.1 for contamination. The concentration of mercury measured at a particular station in a

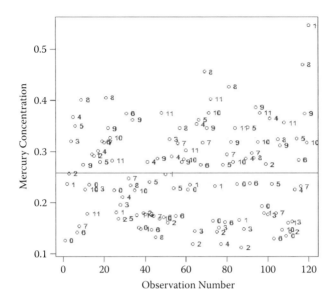

FIGURE 11.1
Index plot of mercury concentrations with study year adjacent to the individual points in the plot.

given year is on the vertical axis, and the observation number is on the horizontal axis. To help display the trends, each data point is coded by the year of the observation (0 to 11 for years 1995 to 2006). The horizontal line on the index plots shows the overall average concentration for all 120 observations. A general observation from the plot is that the majority of the points above the horizontal line correspond to the later years of the study (i.e., years 6–11), indicating there was more contamination toward the end of the study than at the beginning of the study. With this type of plot, it is difficult to explore the data pattern in more detail than this and to make observations about the detail of the trend and whether the same trend is observed at each station.

Box plots showing the distribution of the measurement for each of the 12 study years are shown in Figure 11.2. Mercury levels were certainly trending in a positive direction over time. The spread in the box plots indicates that the mercury measurements were variable from one station to the next, suggesting that some stations were associated with higher levels of contamination than others.

Before introducing various methods for quantifying trend, it is often the case that we would like simply to compare the response across two time points. If our interest is only in how stations changed between 2 years (e.g., the initial and final year of the study), a paired t-test can be used. One proviso is that the data in both groups should be normally distributed, although as long as the data set is large enough the test is robust for departures from normality (Sawilowsky and Blair, 1992). The test statistic is

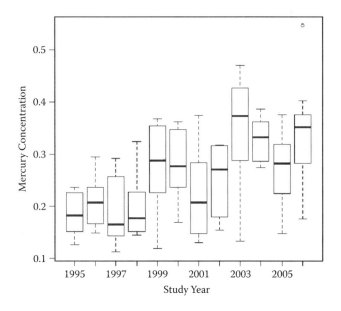

FIGURE 11.2
Side-by-side box plots of mercury concentration levels for which the spread in each box corresponds to the variability in concentrations across the stations for each study year.

$$\frac{\bar{y}_{2006} - \bar{y}_{1995}}{se\left(\bar{y}_{2006} - \bar{y}_{1995}\right)} \sim t_{11},$$

where $\bar{y}_{2006}$ and $\bar{y}_{1995}$ denote the average measurements from 2006 and 1995, respectively; $se\left(\bar{y}_{2006} - \bar{y}_{1995}\right)$ represents the estimated standard error of the difference in the measurements; and $\sim t_{11}$ denotes the fact that the test statistic follows a t distribution with 11 degrees of freedom. The test statistic is $t = 3.81$ with a p value of 0.0021, showing that the average mercury concentration was significantly greater in the final year of the study than it was at the beginning of the study. The observed mean difference in mercury concentration, comparing 2006 to 1995, is 0.1480 ppm.

Another approach for comparing two time points would be to take the differences in the measurements across the time points and then bootstrap the differences to obtain a confidence interval on the difference in average part-per-million levels. The bootstrap approach is a modern nonparametric computational approach and is easily accomplished through a variety of statistical software packages. For more details on this method, see the work of Manly (2007).

At this point, the graphics considered indicate that mercury concentrations appear to be trending upward over time, and the paired t analysis makes it possible to compare the values at any two time points. It is also possible to probe more into the details of the trend to see if, for example, the rate of increase was consistent through time as in a linear trend or whether the rate increased more quickly at the beginning of the study than at the end as in some curved trends. For this, unit or pooled analyses can be used. The unit analysis approach is the subject of the next section.

11.3 Unit Analyses of Trends

As mentioned above, in a unit analysis of trends separate trend analyses are carried out for each of the sampling units. Whenever there is a random sample of units with repeated measures on each unit, there are generally two main scopes of inference: inference on the individual units themselves and inference on the entire study area from which the units were sampled. In unit analysis, interest is in comparing the results of analyses of each individual unit to one another. First, consider the structural form of the trend for each of the units (i.e., whether the trend is linear or nonlinear). Scatter plots are a useful graphical summary to begin. In Figure 11.3, a scatter plot graph with separate panels for each station is provided. Mercury levels appear to be increasing over time at most of the stations, but the details of the trend differ among stations. For example, station 89 showed little change in mercury levels over time, whereas station 90 showed a fairly substantial increase in mercury over the study period. Also, from Figure 11.3 it appears that a linear trend is an appropriate description of the structural change, so that the next step is to fit separate linear regressions for each of the stations.

A linear statistical model relating mercury concentration to study year is given by

$$merc\_conc_{ij} = \beta_{0,i} + \beta_{1,i} year_j + \varepsilon_{ij}, \quad i = 1, 2, \ldots, 10; j = 0, 1, \ldots, 11, \quad (11.1)$$

where $merc\_conc_{ij}$ denotes the measured mercury concentration for station i in study year j; $\beta_{0,i}$ and $\beta_{1,i}$ denote the y intercept and slope, respectively, associated with the regression line for station i; and ε_{ij} denotes the model errors associated with each station. In a linear regression analysis, the model errors are assumed to follow a normal distribution with mean 0 and constant variance σ^2. The use of the subscript i indicates that the model allows for different y intercepts and slopes for each of the stations. Practically, the intercepts for each station represent the mercury concentration in study year 0 or the baseline levels associated with each station. The slopes represent the expected change in concentration on a per annum basis, with positive values representing a positive annual change, negative values representing a negative annual change, and values close to zero representing little to no annual change.

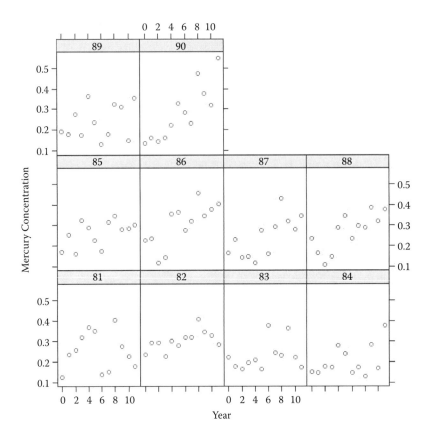

FIGURE 11.3
Scatter plots of mercury concentrations for 12 years of the study for each of the stations (stations are numbered 81–90).

The least-squares fits of the linear regression models for each site are easily obtained using most software packages. Once the regression models are estimated for each station, a display of confidence intervals for the intercepts and slopes associated with the sampling unit can help in making sensible conclusions about the individual units as well as general statements about the population of sampling units. Figure 11.4 shows 95% confidence intervals for the intercepts and slopes (denoted by "year") for each of the 10 stations in the study.

In each panel of Figure 11.4, a vertical line segment extends from zero on the horizontal axis. For every station with the exception of station 82, the mercury concentrations were low at the start of the study period because in the left-hand panel of Figure 11.4 the vertical line represents zero mercury concentration at the start of the study. It appears as if stations 84, 87, and 90 began the study with the lowest levels of contamination, and station 82 began the study with the highest level. In the right-hand panel of Figure 11.4, stations whose corresponding confidence intervals lie to the

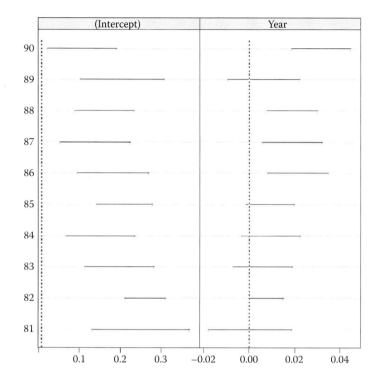

FIGURE 11.4
The 95% confidence intervals on the intercepts and slopes for each of the stations in the mercury concentration data set.

right of the vertical dashed line had statistically significant positive trends over the study period (i.e., stations 82, 86, 87, 88, and 90). The most substantial increase in mercury concentration is observed with station 90, where the estimated slope is $\hat{\beta}_{1,90} = 0.0317$, implying that the mean mercury concentration is estimated to be increasing at a rate of 0.0317 ppm per year. Given this, over a 10-year period the expected increase in concentration at station 90 would be 0.317 ppm.

Although the unit analysis provides direct insight into the trends for each of the individual sampling units, it is useful to make a general statement about the entire study area from which the study sites were selected. To make these types of statements with the unit analysis, summaries of the regression parameter estimates across each of the sampling units can be calculated. As an example, the average trend is estimated to be

$$\bar{\hat{\beta}}_1 = \sum_{i=1}^{10} \hat{\beta}_{1,i} / 10 = 0.0131 \tag{11.2}$$

where $\bar{\hat{\beta}}_1$ denotes the overall average trend across the sample of stations. Thus, on average, for the study area monitored, the mean mercury concentration is estimated to be increasing at a rate of 0.0131 ppm per year, so that for a 10-year period, the expected increase in concentration would be 0.131 ppm. Certainly, there is sampling variability associated with the estimate provided in Equation (11.2), and a 95% confidence interval on this trend could be computed using

$$\bar{\hat{\beta}} \pm t_{8,0.025} s.e.(\bar{\hat{\beta}}) = \bar{\hat{\beta}} \pm 2.306 \frac{sd(\hat{\beta}'_i s)}{\sqrt{10}}$$

$$= 0.0131 \pm 2.306 \frac{0.0094}{\sqrt{10}} \qquad (11.3)$$

$$= 0.01311 \pm 0.0069$$

where $sd(\hat{\beta}'_i s)$ denotes the standard deviation of the sample of observed slope coefficients associated with the 10 stations. In summary, it is estimated that the expected per annum increase in mercury concentration across the entire study area is between 0.0062 and 0.0200 ppm. Note from Equation (11.3) that inference about the entire study area involves a t statistic with 8 degrees of freedom. The 8 degrees of freedom are calculated as 10 – 2 because there are 10 sampling units and 2 regression parameters.

It should be noted that the data from each of the stations are not independent in this study because each station was repeatedly surveyed, although there is independence among the sampling units because the units were randomly selected. If the data were analyzed as if there are $n - 2$ degrees of freedom and the repeated measures structure of the data were ignored, the 120 observations would be pooled and one regression model fitted as

$$merc\_conc_{ij} = \beta_0 + \beta_1 year_j + \varepsilon_{ij}. \qquad (11.4)$$

In Equation (11.4), there is a common population intercept β_0 and slope β_1; the resulting analysis is given in Table 11.1. The overall estimates of the population intercept and slope are identical to the sample means of the estimated intercepts and slopes, respectively, from the unit analysis, but the reported standard errors, t values, and p values are all based on an assumed sample size of 120 instead of the true sample size of 10. When this sort of mistake is made, confidence intervals will be much narrower than they should be, and there will be a high type I error rate when reporting results (i.e., your chance of concluding effects to be statistically significant when they are not truly significant will be much higher than your stated value of α).

TABLE 11.1

Summary from the Analysis of the Model in Equation (11.4) that Falsely Assumes that All 120 Observations in the Data Set Are Independent

Coefficients:

	Estimate	Std. Error	*t* value	Pr(>|*t*|)
(Intercept)	0.18641	**0.01334**	**13.97**	**< 2e-16*****
Year	0.01311	**0.00205**	**6.38**	**3.6e-09*****

—

Significant codes: 0 '***' 0.001 '**' 0.01 '*' 0.05 '.' 0.1 ' ' 1

Residual standard error: **0.0777 on 118 degrees of freedom**

Multiple R squared: 0.256; adjusted R squared: 0.25

F-statistic: 40.7 on 1 and 118 df, *p* value: 3.62e-09

Note: Everything in **bold** is wrongly reported according to the sampling design.

11.4 Pooled Analysis of Trends

The previous section considered analyses that could be used to make inferences about individual sampling units as well as about the population as a whole from which the sampling units were selected. Before beginning the discussion of the pooled analysis of trends, it is important first to note that the conclusions based on the unit analyses will hold true in the pooled analysis. The pooled analysis makes use of a more sophisticated model, but the sophistication of the model does not change the story that is told; it merely tells the story in a different manner.

There were two types of regression model estimated in the previous section: the site-specific regression models as shown in Equation (11.1) and a population-averaged model as shown in Equation (11.4). To start the discussion of the pooled analysis, it is helpful to overlay the fits of these models for each of the stations, as shown in Figure 11.5. In thinking about why time trends differ from station to station, differences caused by natural fluctuations in the lake bed structure, fish diversity, and so on can be considered. As such, the different trend lines associated with each station can be viewed as random fluctuations from the overall average trend line (i.e., the trend line pictured with a solid line in each of the panels in Figure 11.5).

Algebraically, a line is characterized by its slope and *y* intercept; hence, it is reasonable to view each of the station-specific intercepts and slopes as random fluctuations from the population-averaged intercept and slope. Exploiting this notion of randomly varying intercepts and slopes, the pooled analysis makes use of what is commonly referred to as a mixed model. The mixed model for the mercury data would be written as

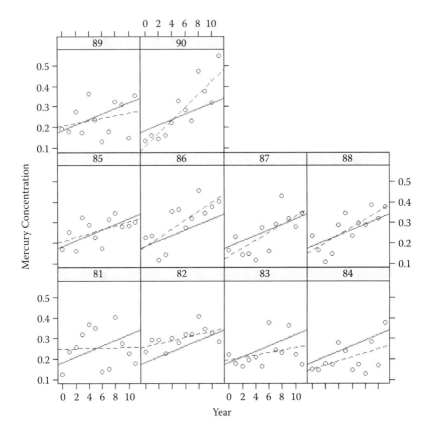

FIGURE 11.5
Observation-specific and population-averaged fits, dashed and solid lines, respectively, of mercury concentration by study year.

$$merc\_conc_{ij} = \beta_0 + \delta_{0,i} + \beta_1 year_j + \delta_{1,i} year_j + \varepsilon_{ij}$$
$$= (\beta_0 + \delta_{0,i}) + (\beta_1 + \delta_{1,i}) year_j + \varepsilon_{ij}$$

(11.5)

where in Equation (11.5) β_0 represents the population-averaged intercept and $\delta_{0,i}$ denotes the random fluctuation from β_0 associated with the ith station. Basically, β_0 denotes the average mercury concentration for the entire study area in the initial year of monitoring and $\delta_{0,i}$ denotes the deviation from this population-averaged value for the ith station. Similarly, β_1 in the equation denotes the expected yearly change in mercury concentration for the entire study area, whereas $\delta_{1,i}$ denotes how the yearly change associated with station i differs from the population average change. It is generally assumed for linear mixed models that the $\delta_{0,i}$ follow independent and identically distributed normal distributions with mean 0 and variance $\sigma_{\delta_0}^2$. The same is also assumed for the random slopes, but the variance is given by $\sigma_{\delta_1}^2$. The model errors are also assumed to follow a normal distribution with mean 0 and variance σ_ε^2. One last assumption is that the random intercepts and slopes are correlated, and the correlation parameter is generally denoted by ρ_{δ_0,δ_1}. For this study, if $\rho_{\delta_0,\delta_1} > 0$, this would imply that those stations with higher levels of mercury at the beginning of the study (quantified by the $\delta_{0,i}$) would be associated with larger yearly changes (quantified by the $\delta_{1,i}$). Conversely, if $\rho_{\delta_0,\delta_1} < 0$, then those stations with lower mercury levels at the beginning of the study would be associated with larger yearly changes.

The reason for the mixed-model terminology for Equation (11.5) is the existence of both fixed and random effects in the model. The fixed effects in the equation are the population-averaged terms β_0 and β_1 because there is a fixed unknown value associated with the mean mercury concentration in the initial year of the study (i.e., β_0) and a fixed, unknown yearly trend associated with the study area as a whole (i.e., β_1). The random effects in the equation are the sampling-unit-specific terms $\delta_{0,i}$ and $\delta_{1,i}$. Because these terms are sampling unit specific and the sampling units were randomly chosen locations from the entire study area, they are called random effects.

The model in Equation (11.5) is generally referred to as a linear mixed model because all the regression parameters (i.e., the $\beta's$ and $\delta's$) enter the model in a linear fashion. It is important to note that there are three sources of variation specified in the equation. These are the variation in the random intercepts $\sigma_{\delta_0}^2$, the variation in the random slopes $\sigma_{\delta_1}^2$, and the residual model variation σ_ε^2. The residual model variation can be thought of as the within-sampling-unit variation. To understand this within-sampling-unit variation, suppose a regression model is estimated only for the data associated with station 90, as shown in Figure 11.6. Every point in Figure 11.6 is assumed to be a realization from a normal distribution whose variation is quantified by σ_ε^2, the within-sampling-unit variation. As such, the normal distribution pictured in Figure 11.6 has variance σ_ε^2. Thus, for a given sampling unit (say station 90) and a given year (say 2002), we expect to observe any mercury concentration values to be within the range of the pictured normal curve. Note that in the data set the mercury concentration in 2002 is considered to be lower than expected.

Each of the variance component estimates can be calculated when fitting a linear mixed model. This analysis can begin by first fitting what is known as the random intercept model

$$merc\_conc_{ij} = (\beta_0 + \delta_{0,i}) + \beta_1 year_j + \varepsilon_{ij} \qquad (11.6)$$

and comparing the model in Equation (11.6) with the one in Equation (11.5). The model in Equation (11.6) only accommodates separate intercepts for each sampling unit, whereas Equation (11.5) accommodates separate slopes and separate intercepts. The analysis of the model in Equation (11.6) is provided

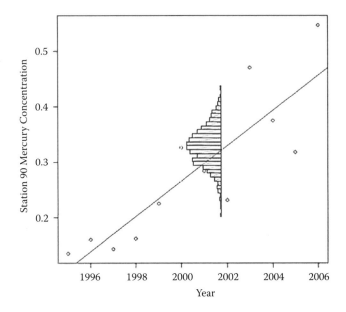

FIGURE 11.6
Regression fit for station 90 with a normal distribution overlaid for the year 2002 demonstrating the interpretation of the within-unit variation.

in Table 11.2, but before interpreting the output in Table 11.2, consider the analysis for the Equation (11.5) model in Table 11.3, which accommodates both random intercepts and random slopes.

In Tables 11.2 and 11.3, the first section of output is a section that provides an indication of the model fit. Akaike's information criterion (AIC) is widely accepted as a model comparison statistic, and lower values of AIC are associated with better-fitting models. Comparing the AICs of the random intercept and random coefficient models (Tables 11.2 and 11.3, respectively), there is a slightly better fit with the random coefficient model. A difference in AIC of less than 2 units is generally taken to indicate models of similar fit. The next section of output provides summaries of the variance components analysis, and the header for this section is "Random Effects." When fitting the random coefficient model, the estimated standard deviation among the intercepts is $\hat{\sigma}_{\delta_0} = 0.0294$, the estimated standard deviation among the random slopes is $\hat{\sigma}_{\delta_1} = 0.0074$, and the estimated residual standard deviation is $\hat{\sigma}_\varepsilon = 0.0703$.

One additional estimate in the "Random Effects" section of Table 11.3 is the correlation between the random intercepts and the random slopes, given by $\hat{\rho}_{\delta_0,\delta_1} = -0.8110$. This negative correlation suggests that those stations that began with lower mercury levels tended to have more substantial yearly changes during the study period. This can be seen by plotting the estimated random intercepts against the estimated random slopes as shown in Figure 11.7.

TABLE 11.2

Computer Output from the Linear Mixed-Effects Analysis of the Random Intercept Model

Linear Mixed-Effects Model Fit

	AIC	BIC	logLik
	−251	−240	129

Random Effects
  Formula: ~1 | Station

	(Intercept)	Residual
Standard Deviation:	0.0222	0.0747

Fixed Effects: merc_conc ~ year

	Value	Standard Error	t Value	p Value
(Intercept)	0.1864	0.01463	12.74	0
Year	0.0131	0.00198	6.63	0

TABLE 11.3

Computer Output from the Linear Mixed-Effects Analysis of the Random Coefficients Model

Linear Mixed-Effects Model Fit

	AIC	BIC	logLik
	−253.164	−236.54	132.582

Random Effects
  Formula: ~year | Station
  Structure: General positive-definite, Log-Cholesky parameterization

	StdDev	Corr
(Intercept)	0.02939642	(Intr)
Year	0.00735664	−0.8220
Residual	0.07032382	

Fixed Effects: merc_conc ~ year

	Value	Standard Error	t Value	p Value
(Intercept)	0.1864092	0.01523949	12.23199	0
Year	0.0121057	0.00297831	4.40038	0
Correlation:				
(Intr)				
Year	−0.811			

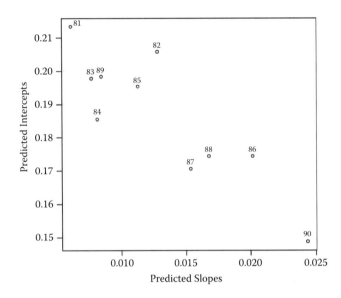

FIGURE 11.7
A scatter plot of the predicted intercepts $\hat{\delta}_{0,i}$ versus the predicted slopes $\hat{\delta}_{1,i}$ for the study data. The negative trend is consistent with $\hat{\rho}_{\delta_0,\delta_1} = -0.8220$ in Table 11.3.

The total estimated variation in the response, $merc\_conc_{ij}$, from Equation (11.5) is given by

$$\text{Var}(merc\_conc_{ij}) = \hat{\sigma}_{\delta_0}^2 + 2\sigma_{\delta_0,\delta_1} year_j + \hat{\sigma}_{\delta_1}^2 year_j^2 + \hat{\sigma}_{\varepsilon}^2. \tag{11.7}$$

See the work of Myers et al. (2010) for more details on this calculation. Regarding notation, $\hat{\sigma}_{\delta_0,\delta_1} = \hat{\rho}_{\delta_0,\delta_1} \cdot \hat{\sigma}_{\delta_0} \cdot \hat{\sigma}_{\delta_1}$, and using this expression, the estimated variance for any station during the initial study year is given by

$$\text{Var}(merc\_conc_{ij}|year=0) = \hat{\sigma}_{\delta_0}^2 + 2\hat{\sigma}_{\delta_0,\delta_1} \cdot 0 + \hat{\sigma}_{\delta_1}^2 \cdot 0 + \hat{\sigma}_{\varepsilon}^2$$

$$= 0.0294^2 + 0.0703^2 = 0.0058 \tag{11.8}$$

where $\hat{\sigma}_{\delta_0,\delta_1}$ is calculated as $\hat{\sigma}_{\delta_0,\delta_1} = -0.8110 \cdot 0.0294 \cdot 0.0074$. One of the uses for the estimated variance in the mercury concentrations for stations is to discern how much of the response variation is attributable to variation across the study site (i.e., variation among stations, δ_{0i} and δ_{1i}) and how much of the variation is caused by the within-site variation (i.e., variation within stations σ_{ε}^2).

The proportion of response variation attributable to differences among stations (sites) for a given year is

$$Proportion\ due\ to\ stations = \frac{\hat{\sigma}_{\delta_0}^2 + 2\hat{\sigma}_{\delta_0,\delta_1} year_j + \hat{\sigma}_{\delta_1}^2 year_j^2}{\hat{\sigma}_{\delta_0}^2 + 2\hat{\sigma}_{\delta_0,\delta_1} year_j + \hat{\sigma}_{\delta_1}^2 year_j^2 + \hat{\sigma}_{\varepsilon}^2}. \tag{11.9}$$

For example, in the initial year of the study (i.e., $year_j = 0$), the estimated proportion of response variation caused by site-to-site differences is

$$\frac{0.0294^2}{(0.0294^2 + 0.0703^2)} = 0.1487.$$

Knowing the proportions of variation that are attributable to site-to-site differences versus within-site variation is helpful in planning future allocation of sampling effort.

Recall that one of the major goals in trend analysis is to describe the overall trend for a population as well as the individual observed trends associated with each sampling unit. From the model given in Equation (11.5), the predicted mercury concentration for the ith station is given by

$$merc\_conc_{ij} = (\hat{\beta}_0 + \hat{\delta}_{0,i}) + (\hat{\beta}_1 + \hat{\delta}_{1,i})year_j. \tag{11.10}$$

The ^ notation in Equation (11.10) suggests that replacing the unknown parameters from Equation (11.5) with their estimated values from the analysis. Recall that the values of $\hat{\beta}_0$ and $\hat{\beta}_1$ represent the estimated mercury concentration in the initial study year and the yearly rates of change, respectively, associated with the entire study area (i.e., the population from which the sampling units were drawn). The values of $\hat{\beta}_0$ and $\hat{\beta}_1$ for the mercury concentration data are found in Table 11.3 in the section "Fixed Effects" with $\hat{\beta}_0 = 0.1864$ and $\hat{\beta}_1 = 0.0131$. Note that these estimates are precisely the same estimates observed when taking the sample means of the unit intercepts and slopes in Equation (11.2). Also, note that the estimated standard error of $\hat{\beta}_1$ is reported to be 0.0030, and that this value is the same standard error computed for $\hat{\beta}_1$ in Equation (11.3) when conducting the unit analysis.

To obtain the individual prediction equations for each sampling unit requires these predicted values of $\hat{\beta}_0$ and $\hat{\beta}_1$ and the predicted values of $\hat{\delta}_{i,0}$ and $\hat{\delta}_{i,1}$. The predicted trend equation for station 90 is given by

$$merc\_conc_{90} = (\hat{\beta}_0 + \hat{\delta}_{0,1=90}) + (\hat{\beta}_1 + \hat{\delta}_{1,i=90})year,$$

and the predicted trend equation for the station with the least trend (station 81) is given by

$$merc\_conc_{81} = (\hat{\beta}_0 + \hat{\delta}_{0,i=81}) + (\hat{\beta}_1 + \hat{\delta}_{1,i=81})year.$$

With the information from Table 11.4,

$$merc\_conc_{90} = 0.1488 + 0.0244 \ year,$$

TABLE 11.4

Predicted Intercepts and Slope Deviations from the
Linear Mixed-Model Random Coefficient Model of the
Mercury Concentration Data

Station	Intercept	Slope
81	0.2133	0.0061
82	0.2056	0.0128
83	0.1977	0.0077
84	0.1856	0.0081
85	0.1953	0.0113
86	0.1744	0.0201
87	0.1707	0.0154
88	0.1745	0.0168
89	0.1982	0.0085
90	0.1488	0.0244

and

$$merc\_conc_{81} = 0.2133 + 0.0061\, year.$$

Hence, the yearly expected change in mercury concentration ranged from
0.0061 to 0.0244 ppm across the study area.

11.5 Checking for Model Adequacy

As with any model-fitting analysis, it is important to check that the proposed
model is a reasonable approximation to the underlying true population process, and that no assumptions used in the analysis are violated.

One of the ways to review the model for adequacy is to look at the model
residuals. For a linear regression model, as used in this example, the model
residuals should appear to be normally distributed. However, because of the
use of a linear mixed model, there are two other components of the model that
need to be checked for normality. These are the random slopes and the random
intercepts. Normal Q-Q plots of the predicted coefficients for the slope and the
intercept are shown in Figure 11.8. In both graphs, the line appears to be straight
so that the normality assumptions for the intercept and slope are satisfied.

Given the structure of the data, checking for normality in the model residuals involves looking at each set of residuals from each station as shown in
Figure 11.9, where the residuals are displayed in a box plot for each station.
There does not appear to be any reason to be concerned with the assumption
of normality for the residuals.

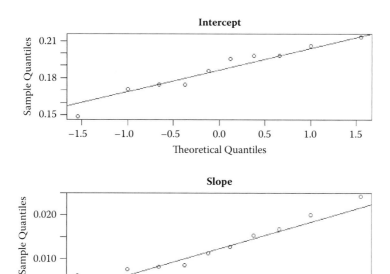

FIGURE 11.8
Normal Q-Q plot of the predicted intercept and slope coefficients.

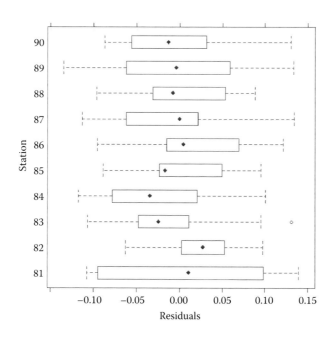

FIGURE 11.9
Model residuals from each station.

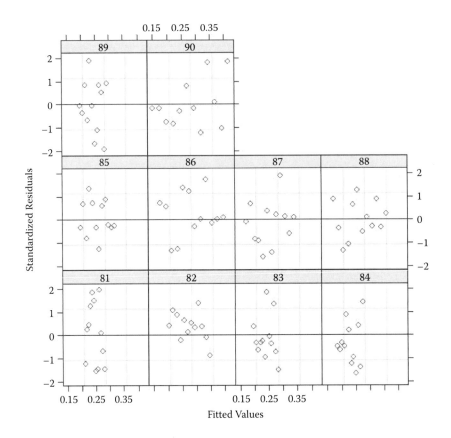

FIGURE 11.10
Standardized residuals from each station plotted against the fitted values.

Another check is to plot the standardized residuals (i.e., the residuals divided by their standard error) versus the fitted values for each station, as shown in Figure 11.10. This is to see if the residuals appear to have a random scatter with no upward or downward trends and if the variation remains equal across the range of fitted values. Departures from a random scatter pattern, for example, with a funnel-shape pattern with more variation with larger fitted values would suggest that the model should be reviewed. If a funnel-shape pattern is observed, then transforming the raw data may be a suitable solution, or a more sophisticated analysis using generalized linear models may be needed. For the present study, there is no indication from looking at the plots in Figure 11.10 that the pattern is not random.

Standardized residuals tend to be used for these types of analyses because this makes it easier to detect any unusual observations. Most of the standardized residuals should be in the range between −3 and 3, and anything outside this should be examined carefully. An outlying data point might be a coding error that can be easily fixed or might suggest some problem with the model.

11.6 Summary

This chapter discussed some ways to assess trend using the mercury concentration in a lake as an example, with measurements from fish tissue of mercury concentration (in ppm) taken each year from 1995 to 2006 from 10 stations. The feature of this data set is that the data were collected from sample units (stations) repeatedly over time, so that they are referred to as repeated measures data.

First, a test for a difference between any two time periods was considered. Then, a unit analysis of trend was carried out with separate analyses for each of the 10 sampling stations using linear regressions. By taking the average for all stations, the mean mercury concentration was estimated with a confidence interval. To be able to make inferences for the entire study area from the sample, a pooled analysis approach was considered. Linear mixed models were used with variation in trend lines for the slope and intercept among the 10 stations explicitly included in the model, with the stations considered as random effects and the year as a fixed factor. It was found that there was some variation among stations in the mercury concentration to begin with as well as with the rate at which levels were increasing, with some stations staying nearly the same. Stations with higher levels of mercury at the beginning tended not to increase at as high a rate as those stations with lower initial levels.

This example data set illustrates the usefulness of the linear mixed model for assessing trend. The example had balanced data with an equal number of repeated observations for each station, but linear mixed models can be used for unbalanced data and with more complex nested experimental designs. An example of a more complex nested design would be if there were surveys over the four seasons within each year, surveys in repeat months within each season, and multiple days of surveying within each month. The model can also be further extended to accommodate data that are not normally distributed; this extension would use a more flexible class of models called generalized linear mixed models (Myers et al., 2010).

References

Abrahamson, I.L., Nelson, C.R., and Affleck, D.L.R. (2011). Assessing the performance of sampling designs for measuring the abundance of understory plants. *Ecological Applications* 21: 452–464.

Acharya, B. Bhattarai, G., de Gier, A., and Stein, A. (2000). Systematic adaptive cluster sampling for the assessment of rare tree species in Nepal. *Forest Ecology and Management* 137: 65–73.

Aerts, R., November, E., Van der Borght, I., Behailu, M., Hermy, M., and Muys, B. (2006). Effects of pioneer shrubs on the recruitment of the fleshy-fruited tree *Olea europaea* ssp. *cuspidata* in Afromontane savanna. *Applied Vegetation Science* 9: 117–126.

Agresti, A. (1994). Simple capture-recapture models permitting unequal catchability and variable sampling effort. *Biometrics* 50: 494–500.

Ahlo, J.M. (1990). Logistic regression in capture-recapture models. *Biometrics* 46: 623–635.

Akaike, H. (1973). Information theory as an extension of the maximum likelihood principle. In: B.N. Petrov and F. Csaki (Eds.), *Second International Symposium on Information Theory*. Akademiai Kiado, Budapest, pp. 267–281.

Amstrup, S.C., McDonald, T.L., and Manly, B.F.J. (2005). *Handbook of Capture–Recapture Analysis*. Princeton University Press, Princeton, NJ and Oxford.

Anderson, D.R. and Pospahala, R.S. (1970). Correction of bias in belt transects of immotile objects. *Journal of Wildlife Management* 34: 141–146.

Anderson, D.R., Burnham, K.P., and White, G.C. (1994). AIC model selection in over-dispersed capture-recapture data. *Ecology* 75: 1780–1793.

Applegate, D.L., Bixby, R.E., Chvátal, V., and Cook, W. (2006). *The Traveling Salesman Problem. A Computational Study*. Princeton University Press, Princeton, NJ.

Arabkhedri, M., Lai, F.S., Noor-Akma, I., and Mohamad-Roslan, M. K. (2010). An application of adaptive cluster sampling for estimating total suspended sediment load. *Hydrology Research* 41: 63–73.

Arnason, A.N., Shar, L., Neilson, D., and Boyer, G. (1998). *RUNPOPAN. Installation and User's Manual for Running POPAN-5 on IBM PC Microcomputers under Windows 3.1/32S, 95, and NT*. Scientific Report. Department of Computer Science, University of Manitoba, Winnipeg, Canada.

Assunção, R.M. (1994). Testing spatial randomness by means of angles. *Biometrics* 50: 531–537.

Assunção, R.M. and Reis, I.A. (2000). Testing spatial randomness: a comparison between T^2 methods and modifications of the angle test. *Brazilian Journal of Probability and Statistics* 14: 71–86.

Bailey, L.L., Simons, T.R., and Pollock, K.H. (2004). Estimating site occupancy and species detection probability parameters for terrestrial salamanders. *Ecological Applications* 14: 692–702.

Baillargeon, S. and Rivest, L.P. (2007). Rcapture: loglinear models for capture-recapture in R. *Journal of Statistical Software*, 19(5).

Baillargeon, S. and Rivest, L.P. (2012). *Rcapture: Loglinear Models for Capture-Recapture Experiments, R Package Version 1.3-1.* CRAN.R-project.org/package=Rcapture.

Barabesi, L. (2001). A design-based approach to the estimation of plant density using point-to-plant sampling. *Journal of Agricultural, Biological, and Environmental Statistics* 6(1): 89–98.

Barbraud, C., Nichols, J.D., Hines, J.E., and Hafner, H. (2003). Estimating rates of local extinction and colonization in colonial species and an extension to the meta-population and community levels. *Oikos* 101: 113–126.

Barlow, J. and Sexton, S. (1996). *The Effect of Diving and Searching Behavior on the Probability of Detecting Track-Line Groups, g_0, of Long-Diving Whales during Line-Transect Surveys.* Administrative Report LJ-96-14. Available from NMFS Southwest Fisheries Science Center, P.O. Box 271, LaJolla, California 92038. 21 pp.

Barrios, B., Arellano, G., and Koptur, S. (2011). The effects of fire and fragmentation on occurrence and flowering of a rare perennial plant. *Plant Ecology* 212: 1057–1067.

Bearer, S., Linderman, M., Huang, J., An, L., He, G., and Liu, J. (2008). Effects of fuel-wood collection and timber harvesting on giant panda habitat use. *Biological Conservation* 141: 385–393.

Besag, J. and Gleaves, J.T. (1973). On the detection of spatial pattern in plant communities. *Bulletin of the International Statistical Institute* 45(1): 153–158.

Borchers, D.L., Zucchini, W., and Fewster, R.M. (1988). Mark-recapture models for line transect surveys. *Biometrics* 54: 1207–1220.

Bostoen, K., Chalabi, Z., and Grais, R.F. (2007). Optimisation of the T-square sampling method to estimate population sizes. *Emerging Themes in Epidemiology* 4(7) (June) (http://www.ete-online.com/content/pdf/1742-7622-4-7.pdf).

Breidt, F.J. (1995). Markov chain designs for one-per-stratum sampling. *Survey Methodology* 21: 63–70.

Brown, J.A. (1999). A comparison of two stratified sampling designs: adaptive cluster sampling and a two-phase sampling design. *Australia and New Zealand Journal of Statistics* 41: 395–404.

Brown, J.A. (2003). Designing an efficient adaptive cluster sample. *Environmental and Ecological Statistics* 10: 95–105.

Brown, J.A. (2011). Adaptive sampling of ecological populations. In: Y. Rong (Ed.), *Environmental Statistics and Data Analysis.* ILM, Hertfordshire, UK, pp. 81–96.

Brown, J.A. and Manly, B.F.J. (1998). Restricted adaptive cluster sampling. *Environmental and Ecological Statistics* 5: 47–62.

Brown, J.A., Salehi, M.M., Moradi, M., Bell, G., and Smith, D.R. (2008). An adaptive two-stage sequential design for sampling rare and clustered populations. *Population Ecology* 50: 239–245.

Brown, J.A., Salehi, M.M., Moradi, M., Panahbehagh, B., and Smith, D.R. (2012). Adaptive survey designs for sampling rare and clustered populations. *Mathematics and Computers in Simulation* (early access online) http://dx.doi.org/10.1016/j.matcom.2012.09.008.

Brown, V., Jacquier, G., Coulombier, D., Balandine, S., Belanger, F., and Legros, D. (2001). Rapid assessment of population size by area sampling in disaster situations. *Disasters* 25: 164–171.

Brownie, C., Anderson, D.R., Burnham, K.P., and Robson, D.S. (1985). *Statistical Inference from Band Recovery Data—A Handbook.* U.S. Fish and Wildlife Service Resource Publication 156. U.S. Fish and Wildlife Service, Washington, DC.

Buckland, S.T., Anderson, D.R., Burnham, K.P., Laake, J.L., Borcher, D.L., and Thomas, L. (2001*). Introduction to Distance Sampling.* Oxford University Press, Oxford.

Buckland, S.T., Anderson, D.R., Burnham, K.P., Laake, J.L., Borchers, D.L., and Thomas, L. (2004). *Advanced Distance Sampling.* Oxford University Press, Oxford.

Buckland, S.T., Laake, J.L., and Borchers, D.L. (2010). Double-observer line transect methods: Levels of independence *Biometrics* 66: 169-177.

Burnham, K.P. and Anderson, D.R. (1992). Data-based selection of an appropriate biological model: the key to modern data analysis. In D.R. McCullogh and R.H. Barrett (Eds.), *Wildlife 2001.* Elsevier Applied Science, London, pp. 16–30.

Burnham, K.P. and Anderson, D.R. (2002). *Model Selection and Multimodel Inference: A Practical Information–Theoretic Approach,* 2nd edition. Springer-Verlag, New York.

Burnham, K.P. and Overton, W.S. (1978). Estimation of the size of a closed population when capture probabilities vary among animals. *Biometrika* 65: 625–633.

Burnham, K.P., Anderson, D.R., and Laake, J.L. (1980). Estimation of density from line transect sampling of biological populations. *Wildlife Monographs* 72: 1–202.

Burnham, K.P., Anderson, D.R., and White, G.C. (1994). Evaluation of the Kullback-Leibler discrepancy for model selection in open population capture-recapture models. *Biometrical Journal* 36: 299–315.

Burnham, K.P., White, G.C., and Anderson, D.R. (1995). Model selection in the analysis of capture-recapture data. *Biometrics* 51: 888–898.

Burnham, K.P., Anderson, D.R., White, G.C., Brownie, C., and Pollock, K.H. (1987). *Design and Analysis Methods for Fish Survival Experiments Based on Release-Recapture.* American Fisheries Society, Bethesda, MD.

Byers, J.E., Reichard, S., Randall, J.M., Parker, I.M., Smith, C.S. Lonsdale, W.M., Atkinson, I.A. E., Seastedt, T.R., Williamson, M., Chornesky, E., and Hayes, D. (2002). Directing research to reduce the impacts of nonindigenous species. *Conservation Biology* 16: 630–640.

Byth, K. (1982). On robust distance-based intensity estimators. *Biometrics* 38: 127–135.

Casella, R. and Berger, R.L. 2001. *Statistical Inference,* 2nd edition. Pacific Grove, Duxbury, CA.

Catana, A.J. (1963). The wandering quarter method of estimating population density. *Ecology* 44: 349–360.

Chaloner, K. and Verdinelli, I. (1995). Bayesian experimental design: a review. *Statistical Science* 10: 273–304.

Chao, A. (1987). Estimating the population size for capture-recapture data with unequal catchability. *Biometrics* 43: 783–791.

Chao, A. and Huggins, R.M. (2005a). Classical closed-population capture-recapture models. In: S.C. Amstrup, T.L. McDonald, and B.F.J. Manly (Eds.), *Handbook of Capture-Recapture Analysis.* Princeton University Press, Princeton, NJ, pp. 58–87.

Chao, A. and Huggins, R.M. (2005b). Modern closed-population capture-recapture models. In: S.C. Amstrup, T.L. McDonald, and B.F.J. Manly (Eds.), *Handbook of Capture–Recapture Analysis.* Princeton University Press, Princeton, NJ, pp. 22–35.

Chao, A. and Yang, H.C. (2003). *Program CARE-2 (for Capture-Recapture Part 2). Program and User's Guide.* Available at http//chao.stat.nthu.edu.tw.

Chao, A., Yip, P.S.F., Lee, S.-M., and Chu, W. (2001). Population size estimation based on estimating functions for closed capture-recapture models. *Journal of Statistical Planning and Inference* 92: 213–232.

Chapman, D.G. (1951). Some properties of the hypergeometric distribution with applications to zoological sample censuses. *University of California Publications in Statistics* 1: 131–160.

Christman, M.C. and Lan, F. (2001). Inverse adaptive cluster sampling. *Biometrics* 57: 1096–1105.

Cochran, W.G. (1977) *Sampling Techniques*, 3rd edition, Wiley, New York.

Coggins, S.B., Coops, N.C., and Wulder, M.A. (2011). Estimates of bark beetle infestation expansion factors with adaptive cluster sampling. *International Journal of Pest Management* 57: 11–21.

Conners, M.E. and Schwager, S.J. (2002). The use of adaptive cluster sampling for hydroacoustic surveys. *Journal of Marine Science: Journal du Conseil* 59: 1314–1325.

Cormack, R.M. (1964). Estimates of survival from the sighting of marked animals. *Biometrika* 51: 429–438.

Cormack, R.M. (1989). Loglinear models for capture-recapture. *Biometrics* 45: 395–413.

Cottam, G. (1947). A point method for making rapid surveys of woodlands (Abstract). *Bulletin of the Ecological Society of America* 28: 60.

Cottam, G., Curtis, J.T., and Catana, A.J., Jr. (1957). Some sampling characteristics of a series of aggregated populations. *Ecology* 38: 610–622.

Cottam, G., Curtis, J.T., and Hale, B.W. (1953). Some sampling characteristics of a population of randomly dispersed individuals. *Ecology* 34: 741–757.

Davis J.G. and Smith, D.D. (2011). Testing the utility of an adaptive cluster sampling method for monitoring a rare and imperiled darter. *North American Journal of Fisheries Management* 31:1123–1132.

Díaz-Gamboa, R. (2009). Relaciones tróficas de los cetáceos teutófagos con el calamar gigante *Dosidicus gigas* en el Golfo de California. PhD dissertation [in Spanish]. CICIMAR IPN, México. 103 pp.

Diggle, P.J. (2003). *Statistical Analysis of Spatial Point Patterns*, 2nd edition. Arnold, London.

Dixon, W.J. and Massey, F.J. (1983). *Introduction to Statistical Analysis*. McGraw-Hill, New York.

Drummer, T.D., Degange, A.R., Pank, L.L., and McDonald, L.L. (1990). Adjusting for group size influence in line transect sampling. *Journal of Wildlife Management* 54: 511–514.

Drummer, T.D. and McDonald, L.L. (1987). Size bias in line transect sampling. *Biometrics* 43: 13–21.

Edwards, D. (1998). Issues and themes for natural resources trend and change detection. *Ecological Applications* 8: 323–325.

Efford, M.G. (2012). *DENSITY 5.0: Software for Spatially Explicit Capture-Recapture*. Department of Mathematics and Statistics, University of Otago, Dunedin, New Zealand. Available at http://www.otago.ac.nz/density.

Efford, M.G., Dawson, D.K., and Robbins, C.S. (2004). DENSITY: software for analysing capture-recapture data from passive detector arrays. *Animal Biodiversity and Conservation* 27: 217–228.

Efford, M.G. and Fewster, R.M. (2013). Estimating population size by spatially explicit capture-recapture. *Oikos* 122: 918–928.

Efron, B. and Tibshirani, R.J. (1993). *An Introduction to the Bootstrap*. Chapman and Hall, New York.

Engeman, R.M., Sugihara, R.T., Pank, L.F., and Dusenberry, W.E. (1994). A comparison of plotless density estimators using Monte Carlo simulation. *Ecology* 75: 1769–1779.

ErfaniFard, Y., Fegghi, J. Zobeiri, M., and Namiranian, M. (2008). Comparison of two distance methods for forest spatial pattern analysis (case study: Zagros Forests of Iran). *Journal of Applied Sciences* 8: 152–157.

Exeter Software. (2009). *Ecological Methodology, Version 7.0*. Exeter Software, Setauket, NY.

Fancy, S.G., Gross, J.E., and Carter, S.L. (2009). Monitoring the condition of natural resources in U.S. national parks. *Environmental Monitoring and Assessment* 151: 161–174.

Fewster, R.M, Buckland, S.T, Burnham, K.P., Borchers, D.L., Jupp, P.E., Laake, J.L., and Thomas, L. (2009). Estimating the encounter rate variance in distance sampling. *Biometrics* 65: 225–236.

Fisher, R.A. and Ford, E.B. (1947). The spread of a gene in natural conditions in a colony of the moth *Panaxia dominula*. *Heredity* 1: 143–174.

Fiske, I. and Chandler, R.B. (2011). Unmarked: an R package for fitting hierarchical models of wildlife occurrence and abundance. *Journal of Statistical Software* 43: 1–23 (http://www.jstatsoft.org/v43/i10/).

Francis, R.I.C.C. (1984). An adaptive strategy for stratified random trawl surveys. *New Zealand Journal of Marine and Freshwater Research* 18: 59–71.

Gattone, S.A. and Di Battista, T. (2011). Adaptive cluster sampling with a data driver stopping rule. *Statistical Methods and Applications* 20: 1–21.

Gerrard, D.J. (1969). *Competition Quotient: A New Measure of the Competition Affecting Individual Forest Trees*. Research Bulletin 20. Agricultural Experiment Station, Michigan State University, East Lansing.

Goldberg, N.A., Heine, J.N., and Brown, J.A. (2007). The application of adaptive cluster sampling for rare subtidal macroalgae. *Marine Biology* 151: 1343–1348.

Grais, R.F., Coulombier, D., Ampuero, J., Lucas, M.E.S., Barretto, A.T., Jacquier, G., Diaz, F., Balandine, S., Mahoudeau, C., and Brown, V. (2006). Are rapid population estimates accurate? A field trial of two different assessment methods. *Disasters* 30: 364–376.

Guillera-Arroita, G., Ridout, M.S., and Morgan, B.J.T. (2010). Design of occupancy studies with imperfect detection. *Methods in Ecology and Evolution* 1: 131–139.

Hall, P., Melville, G., and Welsh, A.H. (2001). Bias correction and bootstrap methods for a spatial sampling scheme. *Bernoulli* 7: 829–846.

Halton, J.H. (1960). On the efficiency of certain quasi-random sequences of points in evaluating multi-dimensional integrals. *Numerische Mathematik* 2: 84–90.

Hanselman, D.H., Quinn, T.J., Lunsford, C., Heifetz, J., and Clausen, D. (2003). Applications in adaptive cluster sampling of Gulf of Alaska rockfish. *Fisheries Bulletin* 101: 501–513.

Harbitz, A., Ona, E., and Pennington, M. (2009). The use of an adaptive acoustic-survey design to estimate the abundance of highly skewed fish populations. *ICES Journal of Marine Science* 66: 1349–1354.

Hayne, D.W. (1949). An examination of the strip census method for estimating animal populations. *Journal of Wildlife Management* 13: 145–157.

Henderson, A. (2009). Using the *T*-square sampling method to estimate population size, demographics and other characteristics in emergency food security assessments (EFSAs). Emergency Food Security Assessments (EFSAs), Technical guidance sheet no. 11. World Food Programme, Rome.

Hines, W.G.S. and O'Hara Hines, R.J. (1979). The Eberhardt statistic and the detection of nonrandomness of spatial point distributions. *Biometrika* 66: 73–79.

Hornbach, D.L., Hove, M.C., Dickinson, B.D., MacGregor, K.R., and Medland, K.R. (2010). Estimating population size and habitat associations of two federally endangered mussels in the St. Croix River, Minnesota and Wisconsin, USA. *Aquatic Conservation: Marine and Freshwater Ecosystems* 20: 250–260.

Horvitz, D.G. and Thompson, D.J. (1952). A generalization of sampling without replacement from a finite universe. *Journal of the American Statistical Association* 47: 663–685.

Huggins, R.M. (1989). On the statistical analysis of capture experiments. *Biometrika* 76: 133–140.

Huggins, R.M. (1991). Some practical aspects of a conditional likelihood approach to capture-recapture experiments. *Biometrics* 47: 725–732.

Huggins, R. and Hwang, W. (2011). A review of the use of conditional likelihood in capture-recapture experiments. *International Statistical Review* 79: 385–400.

Jackson, C.H.N. (1939). The analysis of an animal population. *Journal of Animal Ecology* 8: 238–246.

Jackson, C.H.N. (1940). The analysis of a tsetse fly population: I. *Annals of Eugenics* 10: 332–369.

Jackson, C.H.N. (1944). The analysis of a tsetse fly population: II. *Annals of Eugenics* 12: 176–205.

Jackson, C.H.N. (1948). The analysis of a tsetse fly population: III. *Annals of Eugenics* 14: 91–108.

Jolly, G.M. (1965). Explicit estimates from capture-recapture data with both death and immigration—stochastic model. *Biometrika* 52: 225–247.

Jolly, G.M. and Hampton, I. (1990). A stratified random transect design for acoustic surveys of fish stocks. *Canadian Journal of Fisheries and Aquatic Sciences* 4: 1282–1291.

Kelker, G.H. (1940). Estimating deer populations from a differential hunting loss in the two sexes. *Proceedings of the Utah Academy of Sciences, Arts and Letters* 17: 6–69.

Kelker, G.H. (1944). Sex ratio equations and formulas for determining wildlife populations. *Proceedings of the Utah Academy of Sciences, Arts and Letters* 19–20: 189–198.

Kendall, W.L. and White, G.C. (2009). A cautionary note on substituting spatial subunits for repeated temporal sampling in studies of site occupancy. *Journal of Applied Ecology* 46: 1182–1188.

Kenkel, N.C., Juhász-Nagy, P., and Podani, J. (1989). On sampling procedures in population and community ecology. *Vegetatio* 83: 195–207.

Kitanidis, P.K. (1997). *Introduction to Geostatistics: Applications in Hydrogeology*. Cambridge University Press, Cambridge.

Kleinn, C. and Vilčko, F. (2006). A new empirical approach for estimation in k-tree sampling. *Forest Ecology and Management* 237: 522–533

Krebs, C.J. (1999). *Ecological Methodology*, 2nd edition. Benjamin Cummings, Menlo Park, CA.

Laake, J., Borchers, D., Thomas, L., Miller D., and Bishop, J. (2013). *MRDS: Mark-Recapture Distance Sampling (mrds)*. R package version 2.1.2. http://CRAN.R-project.org/package=mrds.

Lamacraft, R.R., Friedel, M.H., and Chewings, V.H. (1983). Comparison of distance based density estimates for some arid rangeland vegetation. *Australian Journal of Ecology* 8: 181–187.

Lancia, R.A., Pollock, K.H., Bishir, J.W., and Conner, M.C. (1988). A white-tailed deer harvesting strategy. *Journal of Wildlife Management* 52: 589–595.

Laplace, P.S. (1786). Sur les naissances, les mariages et les morts a Paris, despuis 1771 jusqu'en 1784, et dans toute l'etendue de La France, pendant les anneés 1781 et 1782. *Memoires de L' Academie Royale des Sciences de Paris*. In: *Oeuvres Complétes de Laplace publiees sus les auspices de L'Academie de Sciences par MM. Les Secretaries Perpétuels. Tome Onziéme*. Paris: Gauthier-Villars, 1895, pp. 35–46. ftp://ftp.bnf.fr/007/N0077599_PDF_1_-1DM.pdf.

Lebreton, J.-D., Burnham, K.P., Clobert, J., and Anderson, D.R. (1992). Modeling survival and testing biological hypotheses using marked animals: a unified approach with case studies. *Ecological Monographs* 62: 67–118.

Lebreton, J.-D. and North, P.M. (Eds.). (1993). *Marked Individuals in the Study of Bird Populations*. Birkhauser Verlag, Basel.

Lee, S.M. and Chao, A. (1994). Estimating population size via simple coverage for closed capture-recapture models. *Biometrics* 50: 88–97.

Lele, S.R., Moreno, M., and Bayne, E. (2012). Dealing with detection error in site occupancy surveys: what can we do with a single survey? *Journal of Plant Ecology* 5: 22–31.

Leslie, P.H. and Chitty, D. (1951). The estimation of population parameters from data obtained by means of the capture-recapture method. I. The maximum likelihood equations for estimating the death rate. *Biometrika* 38: 269–292.

Leslie, P.H., Chitty, D., and Chitty, H. (1953). The estimation of population parameters from data obtained by means of the capture-recapture method. III. An example of the practical applications of the method. *Biometrika* 40: 137–169.

Lincoln, F.C. (1930). *Calculating Waterfowl Abundance on the Basis of Banding Returns*. United States Department of Agriculture Circular No. 118. U.S. Department of Agriculture, Washington, DC.

Liu, Y., Chen, Y., Cheng, J., and Lu, J. (2011). An adaptive sampling method based on optimized sampling design for fishery-independent surveys with comparisons with conventional designs. *Fisheries Science* 77: 467–478.

Lo, N.C.H., Griffith, D., and Hunter, J.R. (1997). Using restricted adaptive cluster sampling to estimate Pacific hake larval abundance. *California Cooperative Oceanic Fisheries Investigations* Report 37: 160–174.

Lohr, S.L. (2010). *Sampling: Design and Analysis*, 2d edition. Brooks/Cole, Stamford, CT.

Ludwig, J.A. and Reynolds, J.F. (1988). *Statistical Ecology. A Primer on Methods and Computing*. Wiley, New York.

MacKenzie, D.I., Nichols, J.D., Hines, J.E., Knutson, M.G., and Franklin, A.B. (2003). Estimating site occupancy, colonization, and local extinction when a species is detected imperfectly. *Ecology* 84: 2200–2207.

MacKenzie, D.I., Nichols, J.D., Lachman, G.B., Droege, S., Royle, J.A., and Langtimm, C.A. (2002). Estimating site occupancy rates when detection probabilities are less than one. *Ecology* 83: 2248–2255.

MacKenzie, D.I., Nichols, J.D., Royle, J.A., Pollock, K.H., Bailey, L.L., and Hines, J.E. (2006). *Occupancy Estimation and Modeling: Inferring Patterns and Dynamics of Species Occurrence*. Elsevier, New York.

MacKenzie, D.I., Nichols, J.D., Seamans, M.E., and Gutiérrez, R.J. (2009). Modeling species occurrence dynamics with multiple states and imperfect detection. *Ecology* 90: 823–835.

MacKenzie, D.I. and Royle, J.A. (2005). Designing occupancy studies: general advice and allocating survey effort. *Journal of Applied Ecology* 42: 1105–1114.

Magnussen, S., Kurz, W., Leckie, D.G., and Paradine, D. (2005). Adaptive cluster sampling for estimation of deforestation rates. *European Journal of Forest Research* 124: 207–220.

Manly, B.F.J. (1969). Some properties of a method of estimating the size of mobile animal populations. *Biometrika* 56: 407–410.

Manly, B.F.J. (1984). Obtaining confidence limits on parameters of the Jolly-Seber model for capture-recapture data. *Biometrics* 40: 749–758.

Manly, B.F.J. (2004). Using the bootstrap with two-phase adaptive stratified samples from multiple populations at multiple locations. *Environmental and Ecological Statistics* 11: 367–383.

Manly, B.F.J. (2006). *Randomization, Bootstrap and Monte Carlo Methods in Biology*, 3rd edition. Chapman and Hall/CRC, Boca Raton, FL.

Manly, B.F.J. (2007). *Randomization, Bootstrap, and Monte Carlo Methods in Biology*, 3rd edition. Chapman Hall/CRC, Boca Raton, FL.

Manly, B.F.J. (2009). *Statistics for Environmental Science and Management*, 2nd edition. Chapman and Hall/CRC, Boca Raton, FL.

Manly, B.F.J. and Parr, M.J. (1968). A new method of estimating population size, survivorship and birth rate from capture-recapture data. *Transactions of the Society for British Entomology* 18: 81–89.

Manly, B.F.J., Akroyd, J.M., and Walshe, K.A.R. (2002). Two-phase stratified random surveys on multiple populations at multiple locations. *New Zealand Journal of Marine and Freshwater Research* 36: 581–591.

Manly, B.F.J., Amstrup, S.L., and McDonald, T. L. (2005). Capture-recapture methods in practice. In: S.C. Amstrup, T.L. McDonald, and B.F.J. Manly (Eds.), *Handbook of Capture–recapture Analysis*. Princeton University Press, Princeton, NJ. pp. 266–273.

Manly, B.F.J., McDonald, L.L., and Garner, G.W. (1996). Maximum likelihood estimation for the double count method with independent observers. *Journal of Agricultural, Biological and Environmental Statistics* 1: 170–189.

Manly, B.F.J., McDonald, L.L., and McDonald, T.L. (1999). The robustness of mark-recapture methods: a case study for the northern spotted owl. *Journal of Agricultural, Biological and Environmental Statistics* 4: 78–101.

Manly, B.F.J., McDonald, T.L., Amstrup, S.C., and Regehr, E.V. (2003). Improving size estimates of open animal populations by incorporating information on age. *BioScience* 53: 666–669.

McDonald, L.L. (2004). Sampling rare populations. In: W.L. Thompson (Ed.), *Sampling Rare and Elusive Species*. Island Press, Washington, DC, pp. 11–42.

McDonald, L.L. and Manly, B.F.J. (1989). Calibration of biased sampling procedures. In: L. McDonald, B. Manly, J. Lockwood, and J. Logan (Eds.), *Estimation and Analysis of Insect Populations*. Springer-Verlag, Berlin, pp. 467–483.

McDonald, T. (2012). *mra: Analysis of Mark-Recapture Data,* R package version 2.13. Available at http://CRAN.R-project.org/package=mra.

McDonald, T. (with contributions from Ryan Nielson, James Griswald and Patrick McCann). (2012). *Rdistance: Distance Sampling Analyses.* R package version 1.1. http://CRAN.R-project.org/package=Rdistance.

McDonald, T.L. (2003). Review of environmental monitoring methods: survey designs. *Environmental Monitoring and Assessment* 85: 277–292.

McDonald, T.L. and Amstrup, S.C. (2001). Estimation of population size using open capture-recapture models. *Journal of Agricultural, Biological and Environmental Statistics* 6: 206–220.

McIntyre, G.A. (1952). A method for unbiased selective sampling, using ranked sets. *Australian Journal of Agricultural Research* 3: 385–390.

Mier, K.L. and Picquelle, S.J. (2008). Estimating abundance of spatially aggregated populations: comparing adaptive sampling with other survey designs. *Canadian Journal of Fisheries and Aquatic Sciences* 65: 176–197.

Miller, D. (2013). *Distance: A Simple Way to Fit Detection Functions to Distance Sampling Data and Calculate Abundance/Density for Biological Populations.* R package version 0.7.2. http://CRAN.R-project.org/package=Distance.

Moradi, M. and Salehi, M.M. (2010). An adaptive allocation sampling design for which the conventional stratified estimator is an appropriate estimator. *Journal of Statistical Planning and Inference* 140: 1030–1037.

Morrison, L.W., Smith, D.R., Nichols, D.W., and Young, C.C. (2008). Using computer simulations to evaluate sample design: an example with the Missouri bladderpod. *Population Ecology* 50: 417–425.

Mukhopadhyay, P. (2004). *An Introduction to Estimating Functions.* Alpha Science International, Harrow, UK.

Müller, W.G. (2007). *Collecting Spatial Data,* 3rd edition. Springer, Berlin.

Munholland, P.L. and Borkowski, J.J. (1996). Simple Latin square sampling + 1: a spatial design using quadrats. *Biometrics* 52: 125–136.

Myers, R.H., Montgomery, D.C., Vining, G.G., and Robinson, T.J. (2010). *Generalized Linear Models with Applications in Engineering and the Sciences,* 2nd edition. Wiley, New York.

Nicholls, A.O. (1989). How to make biological surveys go further with generalized linear models. *Biological Conservation* 50: 51–75.

Nichols, J.D., Hines, J.E., Mackenzie, D.I., Seamans, M.E., and Gutiérrez, R.J. (2007). Occupancy estimation and modeling with multiple states and state uncertainty. *Ecology* 88: 1395–1400.

Nielsen, S.E., Haughland, D.L., Bayne, E., and Schieck, J. (2009). Capacity of large-scale, long-term biodiversity monitoring programmes to detect trends in species prevalence. *Biodiversity Conservation* 18: 2961–2978.

Noji, E.K. (2005). Estimating population size in emergencies. *Bulletin of the World Health Organization* 83: 164.

Noon, B.R., Ishwar, N.M., and Vasudevan, K. (2006). Efficiency of adaptive cluster and random sampling in detecting terrestrial herpetofauna in a tropical rainforest. *Wildlife Society Bulletin* 34: 59–68.

Numata, M. (1961). Forest vegetation in the vicinity of Choshi. Coastal flora and vegetation at Choshi, Chiba Prefecture. IV. *Bulletin of Choshi Marine Laboratory, Chiba University* 3: 28–48 (in Japanese).

Olsen, A.R., Sedransk, J., Edwards, D., Gotway, C.A., Liggett, W., Rathbun, S., Reckhow, K.H., and Young, L.J. (1999). Statistical issues for monitoring ecological and natural resources in the United States. *Environmental Monitoring and Assessment* 54: 1–45.

Otis, D.L., Burnham, K.P., White, G.C., and Anderson, D.R. (1978). Statistical inference from capture data on closed animal populations. *Wildlife Monographs*: 62: 3–135.

Outeiro, A., Ondina, P., Fernández, C., Amaro, R., and San Miguel E. (2008). Population density and age structure of the freshwater pearl mussel, *Margaritifera margaritifera*, in two Iberian rivers. *Freshwater Biology* 53: 485–496.

Overton, W.S. and Stehman, S.V. (1995). Design implications of anticipated data uses for comprehensive environmental monitoring programmes. *Environmental and Ecological Statistics* 2: 287–303.

Panahbehagh B., Smith, D.R., Salehi, M.M., Hornbach, D.J., and Brown, J.A. (2011). Multi-species attributes as the condition for adaptive sampling of rare species using two-stage sequential sampling with an auxiliary variable. *International Congress on Modelling and Simulation* (MODSIM), Perth Convention Centre, Australia, December 12–16, pp. 2093–2099.

Petersen, C.G.J. (1895). The yearly immigration of young plaice into the Limfjord from the German Sea. *Report of the Danish Biological Station* 6: 5–84.

Petersen, C.G.J. (1896). The yearly immigration of young plaice into the Limfjiord from the German Sea. *Report of Danish Biological Station* 6: 1–48.

Philippi, T. (2005). Adaptive cluster sampling for estimation of abundances within local populations of low-abundance plants. *Ecology* 86: 1091–1100.

Pollard, J.H., Palka, D., and Buckland, S.T. (2002). Adaptive line transect sampling. *Biometrics* 58: 862–870.

Pollock, K.H. (1991). Modeling capture, recapture, and removal statistics for estimation of demographic parameters for fish and wildlife populations: past, present, and future. *Journal of the American Statistical Association* 86: 225–238.

Pollock, K.H. and Otto, M.C. (1983). Robust estimation of population size in closed animal populations for capture-recapture experiments. *Biometrics* 39: 1035–1049.

Pollock, K.H., Lancia, R.A., Conner, M.C., and Wood, B.L. (1985). A new change-in-ratio procedure robust to unequal catchability of types of animal. *Biometrics* 41: 653–662.

Pollock, K.H., Nichols, J.D., Brownie, C., and Hines, J.E. (1990). Statistical inference for capture-recapture experiments. *Wildlife Monographs* 107: 3–97.

Pradel, R. and Lebreton, J-D. (1991). *User's Manual for Program SURGE Version 4.1.* CEPE/CNRS, Montpellier, France.

Press, W.H., Teukolsky, S.A., Vetterling, W.T., and Flannery, B.P. (1992). *Numerical Recipes in FORTRAN: The Art of Scientific Computing*, 2nd edition. Cambridge University Press, Cambridge.

Quang, P.X. and Lanctot, R.B. (1991). A line transect model for aerial surveys. *Biometrics* 47: 1089–1102.

Quinn, G.P. and Keough, M.J. (2002). *Experimental Design and Data Analysis for Biologists*. Cambridge University Press, Cambridge.

Rasmussen, D.I. and Doman, E.R. (1943). Census methods and their application in the management of mule deer. *Transactions of the North American Wildlife Conference* 8: 369–379.

Rexstad, E. and Burnham, K.P. (1991). *User's Guide for Interactive Program CAPTURE.* Colorado Cooperative Fish and Wildlife Research Unit, Colorado State University, Fort Collins.

Rexstad, E. and Burnham, K.P. (1992). *Users Guide for Interactive Program CAPTURE.* Colorado Cooperative Fish and Wildlife Research Unit, Colorado State University, Fort Collins.

Robertson, B.L., Brown, J.A., McDonald, T.L., and Jakson, P. (2013). BAS: balanced acceptance sampling of natural resources. *Biometrics* 63: 776–784.

Robson, D.S. and Regier, H.A. (1964). Sample size in Petersen mark-recapture experiments. *Transactions of the American Fisheries Society* 93: 215–226.

Royle, J.A. and Link, W.A. (2005). A general class of multinomial mixture models for anuran calling survey data. *Ecology* 86: 2505–2512.

Salehi, M.M. and Brown J.A. (2010). Complete allocation sampling: an efficient and easily implemented adaptive sampling design. *Population Ecology* 52: 451–456.

Salehi, M.M., Moradi, M., Brown, J.A., and Smith, D.R. (2010). Efficient estimators for adaptive stratified sequential sampling. *Journal of Statistical Computation and Simulation* 80: 1163–1179.

Salehi, M.M. and Seber, G.A.F. (1997). Two-stage adaptive cluster sampling. *Biometrics* 53: 959–970.

Salehi, M.M. and Seber, G.A.F. (2002). Unbiased estimators for restricted adaptive cluster sampling. *Australian and New Zealand Journal of Statistics* 44: 63–74.

Salehi, M.M. and Smith, D.R. (2005). Two-stage sequential sampling: a neighborhood-free adaptive sampling procedure. *Journal of Agricultural Biological and Environmental Statistics* 10: 84–103.

Samalens, J.C., Rossi, J.P., Guyn, D., Van Halder, I., Menassieu, P., Piou, D., and Jactel, H. (2007). Adaptive roadside sampling for bark beetle damage assessment. *Forest Ecology and Management* 253: 177–187.

Särndal, C., Swensson, B., and Wretman, J. (1992). *Model Assisted Survey Sampling.* Springer-Verlag, New York.

Sawilowsky, S. and Blair, R.C. (1992). A more realistic look at the robustness and type II error properties of the *t* test to departures from population normality. *Psychological Bulletin* 111: 353–360.

Scheaffer, R.L., Mendenhall, W., and Ott, L. (1979). *Elementary Survey Sampling,* 2nd edition. Duxbury Press, North Scituate, MA.

Scheaffer, R.L., Mendenhall, W., Ott, L., and Gerow, K. (2011). *Elementary Survey Sampling,* 7th edition. PWS-Kent, Boston.

Schnabel, Z.E. (1938). The estimation of the total fish population of a lake. *American Mathematical Monthly* 45: 348–352.

Schreuder, H.T., Gregoire, T.G., and Wood, G.B. (1993). *Sampling Methods for Multiresource Forest Inventory.* Wiley, New York.

Schumacher, F.X. and Eschmeyer, R.W. (1943). The estimation of fish populations in lakes and ponds. *Journal of the Tennessee Academy of Sciences* 18: 228–249.

Schwarz, C.J. and Seber, G.A.F. (1999). Estimating animal abundance: review III. *Statistical Science* 14: 427–456.

Sebastiani, P. and Wynn, H.P. (2000). Maximum entropy sampling and optimal Bayesian experimental design. *Journal of the Royal Statistical Society B* 62: 145–157.

Seber, G.A.F. (1965). A note on the multiple-recapture census. *Biometrika* 52: 249–259.

Seber, G.A.F. (1982). *The Estimation of Animal Abundance and Related Parameters,* 2nd edition. Macmillan, New York.

Seber, G.A.F. (1986). A review of estimating animal abundance. *Biometrics* 42: 267–292.

Seber, G.A.F. (1992). A review of estimating animal abundance II. *International Statistical Review* 60: 129–166.

Seber, G.A.F. and Salehi, M.M. (2004). Adaptive sampling. In: P. Armitage and T. Colton (Eds.), *Encyclopedia of Biostatistics*, Volume 1, 2nd ed. Wiley, Chichester, UK, pp. 59–65.

Sekar, C.C. and Deming, W.E. (1949). On a method of estimating birth and death rates and the extent of registration. *Journal of the American Statistical Association* 44: 101–115.

Shewry, M.C. and Wynn, H.P. (1987). Maximum entropy sampling. *Journal of Applied Statistics* 14: 165–170.

Skalski, J.R. and Millspaugh, J.J. (2006). Application of multidimensional change-in-ratio methods using program USER. *Wildlife Society Bulletin* 34: 433–439.

Skalski, J.R. and Robson, D.S.. (1992). *Techniques for Wildlife Investigations: Design and Analysis of Capture Data*. Academic Press, San Diego, CA.

Smith, D.R., Brown, J.A., and Lo, N.C.H. (2004). Application of adaptive cluster sampling to biological populations. In: W.L. Thompson (Ed.), *Sampling Rare and Elusive Species*. Island Press, Washington, DC, pp. 75–122.

Smith, D.R., Conroy, M.J., and Brakhage, D.H. (1995). Efficiency of adaptive cluster sampling for estimating density of wintering waterfowl. *Biometrics* 51: 777–788.

Smith, D.R., Rogala, J.T., Gray, B.R., Zigler, S.J., and Newton T.J. (2011). Evaluation of single and two-stage adaptive sampling designs for estimation of density and abundance of freshwater mussels in a large river. *River Research Applications* 27: 122–133.

Smith, D.R., Villella, R.F., and Lemarié, D.P. (2003). Application of adaptive cluster sampling to low-density populations of freshwater mussels. *Environmental and Ecological Statistics* 10: 7–15.

Smith, S.J. and Lundy, M.J. (2006). Improving the precision of design-based scallop drag surveys using adaptive allocation methods. *Canadian Journal of Fisheries and Aquatic Sciences* 63: 1639–1646.

Soms, A.P. (1985). Simplified point and interval estimation for removal trapping. *Biometrics* 41: 663–668.

Stanley, T.R. and Burnham, K.P. (1999). A closure test for time-specific capture-recapture data. *Environmental and Ecological Statistics* 6, 197–209.

Stanley, T.R. and Richards, J. (2005). Software review: a program for testing capture-recapture data for closure. *Wildlife Society Bulletin* 33: 782–785.

Steel, R.G.D. and Torrie, J.H. (1980). *Principles and Procedures of Statistics, a Biometrical Approach*. McGraw-Hill Kogakusha, Tokyo.

Steinke, I. and Hennenberg, K.J. (2006). On the power of plotless density estimators for statistical comparisons of plant populations. *Canadian Journal of Botany* 84: 421–432.

Stevens, D.L. (1997). Variable density grid-based sampling designs for continuous spatial populations. *Environmetrics* 8: 167–195.

Stevens, D.L. and Olsen, A.R. (2004). Spatially balanced sampling of natural resources. *Journal of the American Statistical Association* 99: 262–278.

Stuart, A. and Ord, J.K. (1987). *Kendall's Advanced Theory of Statistics*, 5th ed. Griffin and Co, London.

Su, Z. and Quinn, T.J. (2003). Estimator bias and efficiency for adaptive cluster sampling with order statistics and a stopping rule. *Environmental and Ecological Statistics* 10: 17–41.

Sullivan, W.P., Morrison, B.J., and Beamish, F.W.H. (2008). Adaptive cluster sampling: estimating density of spatially autocorrelated larvae of the sea lamprey with improved precision. *Journal of Great Lakes Research* 34: 86–97.

Talvitie, M., Leino, O., and Holopainen, M. (2006). Inventory of sparse forest populations using adaptive cluster sampling. *Silva Fennica* 40: 101–108.

Taylor, M.K., Laake, J., Cluff, H.D., Ramsay, M., and F. Messier. (2002). Managing the risk from hunting for the Viscount Melville Sound polar bear population. *Ursus* 13: 185–202.

Thomas, L., Buckland, S.T., Rexstad, E.A., Laake, J.L., Strindberg, S., Hedley, S.L., Bishop, J.R. B., Marques, T.A., and Burnham, K. P. (2010). Distance software: design and analysis of distance sampling surveys for estimating population size. *Journal of Applied Ecology* 47: 5–14.

Thomas, L., Laake, J.L., Rexstad, E., Strindberg, S., Marques, F.F.C., Buckland, S.T., Borchers, D.L., Anderson, D.R., Burnham, K.P., Burt, M.L., Hedley, S.L., Pollard, J.H., Bishop, J.R.B., and Marques, T.A. (2009). *Distance User's Guide, Distance 6.0. Release 2.* Research Unit for Wildlife Population Assessment, University of St. Andrews, UK. http://www.ruwpa.st-and.ac.uk/distance/.

Thompson, S.K. (1990). Adaptive cluster sampling. *Journal of the American Statistical Association* 85: 1050–1059.

Thompson, S.K. (1991a). Adaptive cluster sampling: designs with primary and secondary units. *Biometrics* 47: 1103–1115.

Thompson, S.K. (1991b). Stratified adaptive cluster sampling. *Biometrika* 78: 389–397.

Thompson, S.K. (2003). Editorial: special issue on adaptive sampling. *Environmental and Ecological Statistics* 10: 5–6.

Thompson, S.K. (2012). *Sampling*, 3rd edition. Wiley, New York.

Thompson, S.K. and Seber, G.A.F. (1996). *Adaptive Sampling*. Wiley, New York.

Tongway, D.J. and Hindley, N.L. (2004). *Landscape Function Analysis: Procedures for Monitoring and Assessing Landscapes (Manual).* CSIRO Sustainable Ecosystems, Canberra.

Trifković, S. and Yamamoto, H. (2008). Indexing of spatial patterns of trees using a mean of angles. *Journal of Forest Research* 13: 117–121.

Turk, P. and Borkowski, J. J. (2005). A review of adaptive cluster sampling: 1990–2003. *Environmental and Ecological Statistics* 12: 55–94.

Udevitz, M.S. and Pollock, K.H. (1991). Change-in-ratio estimators for populations with more than two subclasses. *Biometrics* 47: 1531–1546.

Udevitz, M.S. and Pollock, K.H. (1995). Using effort information with change-in-ratio data for population estimation. *Biometrics* 51: 471–481.

Underwood, A.J. (1997). *Experiments in Ecology: Their Logical Design and Interpretation Using Analysis of Variance.* Cambridge University Press, Cambridge.

Wang, X. and Hickernell, F.J. (2000). Randomized Halton sequences. *Mathematical and Computer Modelling* 32: 887–899.

Wiener, J.G., Evers, D.C., Gay, D.A., Morrison, H.A., and Williams, K.A. (2012). Mercury contamination in the Laurentian Great Lakes region: introduction and overview. *Environmental Pollution* 161: 243–251.

White, G.C. (1996). NOREMARK: population estimation from mark-resighting surveys. *Wildlife Society Bulletin* 24: 50–52.

White, G.C. and Burnham, K.P. (1999). Program MARK: survival estimation from populations of marked animals. *Bird Study* 46 Supplement: 120–138.

White, G.C., Anderson, D.R., Burnham, K.P., and Otis, D.L. (1982). *Capture-Recapture and Removal Methods for Sampling Closed Populations*. Los Alamos National Laboratory Rep. LA-8787-NERP. Los Alamos National Laboratory, Los Alamos, NM.

White, G.C., Burnham, K.P., Otis, D.L., and Anderson, D.R. (1978). *User's Manual for Program CAPTURE*. Utah State University Press, Logan.

White, G.C. and Garrott, R.M. (1990). *Analysis of Wildlife Radio-Tracking Data*. Academic Press, New York.

Williams, B.K., Nichols, J.D., and Conroy, M.J. (2002). *Analysis and Management of Animal Populations*. Elsevier, San Diego, CA.

Yang, H., Kleinn, C., Fehrmann, L., Tang, S., and Magnussen, S. (2011). A new design for sampling with adaptive sampling plots. *Environmental and Ecological Statistics* 18: 223–237.

Yu, H., Jiao, Y., Su, T., and Reid, K. (2012). Performance comparison of traditional sampling designs and adaptive sampling designs for fishery-independent surveys: a simulation study. *Fisheries Research* 113: 173–181.

Zimmermann, D. (1991). Censored distance-based intensity estimation of spatial point processes. *Biometrika* 78: 287–294.

Zippin, C. (1956). An evaluation of the removal method of estimating animal populations. *Biometrics* 12: 163–189.

Zippin, C. (1958). The removal method of population estimation. *Journal of Wildlife Management* 22: 82–90.

Index